Lazarus Fletcher

An Introduction to the Study of Meteorites

With a list of the meteorites represented in the collection

Lazarus Fletcher

An Introduction to the Study of Meteorites
With a list of the meteorites represented in the collection

ISBN/EAN: 9783337218089

Printed in Europe, USA, Canada, Australia, Japan

Cover: Foto ©berggeist007 / pixelio.de

More available books at **www.hansebooks.com**

BRITISH MUSEUM (NATURAL HISTORY)

CROMWELL ROAD, LONDON, S.W.

MINERAL DEPARTMENT.

AN INTRODUCTION

TO THE

STUDY OF METEORITES,

WITH A LIST OF THE METEORITES

REPRESENTED IN THE COLLECTION.

BY

L. FLETCHER, M.A., F.R.S.,

KEEPER OF MINERALS IN THE BRITISH MUSEUM;
FORMERLY FELLOW OF UNIVERSITY COLLEGE AND MILLARD LECTURER AT TRINITY COLLEGE, OXFORD.

[*This Guide-book can be obtained only at the Museum; written applications
should be addressed to " The Director, Natural History Museum,
Cromwell Road, London, S.W."*]

PRINTED BY ORDER OF THE TRUSTEES.

1896.

LONDON:
PRINTED BY WILLIAM CLOWES AND SONS, LIMITED,
STAMFORD STREET AND CHARING CROSS.

PREFACE.

In the accompanying list the geographical arrangement of those meteorites of the fall of which there is no record has been adhered to. This mode of arrangement brings together specimens which have been found in the same district at different times, and may possibly be the result of a single fall. As the dates of discovery or of recognition of meteoric origin, upon which other arrangements are based, are stated very differently in the published lists of the principal meteorite collections, a reference in each instance to the best available report and a brief extract from it are given.

Even as regards the dates of fall of the remaining meteorites there is much discrepancy among the various lists: every case in which the date here given has been found to differ from that recorded in any other list has been verified by reference to reports of the fall.

L. FLETCHER.

November 5th, 1896.

TABLE OF CONTENTS.

———◦◦———

The Meteorites added to the Collection since the issue of the last List (1894) bear the following numbers:—

38, 40, 63, 76, 93, 108, 120, 124, 125, 135, 152, 256, 379, 452-8, 466, 468, 469.

PLAN OF THE MINERAL GALLERY

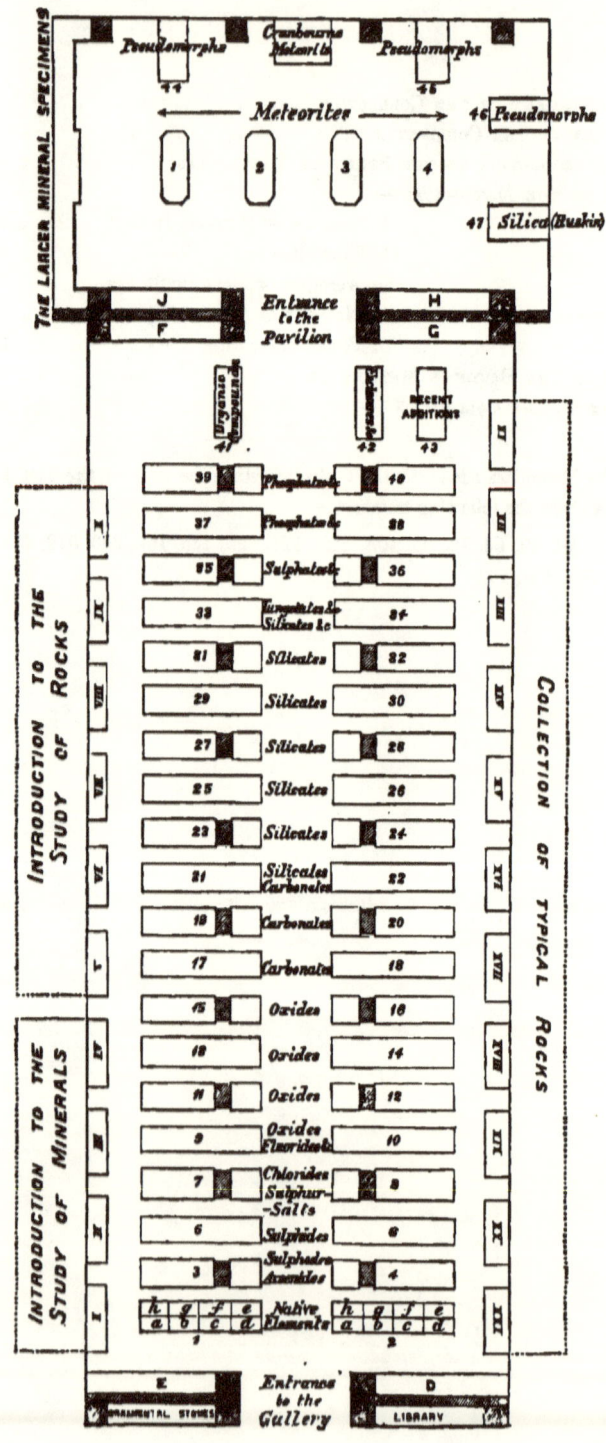

ARRANGEMENT OF THE COLLECTION.

BY ascending the large staircase opposite to the Grand Entrance and turning to the right, the visitor will reach a corridor leading to the Department of Minerals.

From the entrance of the Gallery the large mass of meteoric iron, weighing three and a half tons, found about 1854 at Cranbourne, in Australia, and presented to the Museum in 1862 by James Bruce, Esq., can be seen in the Pavilion at the opposite end of the Gallery.

The other meteorites will be found in the same room, the smaller specimens in the four central cases, and the larger on separate stands. The casts of meteorites are exhibited in the lower parts of the cases.

The specimens referred to in the 'Introduction to the Study of Meteorites' are in case 4, and are arranged, as far as is practicable, in the order of reference.

The remaining specimens are classified as :—

SIDERITES, consisting chiefly of metallic iron (panes 1a–2d) :

SIDEROLITES, consisting chiefly of metallic iron and stony matter, both in large proportions (panes 2e, 2f) : and

AEROLITES, consisting chiefly of stony matter (panes 2g–3n).

At the beginning of each class are placed those meteorites of which the fall has been observed.

The position of any meteorite in the cases may be found by reference to the Index and to the second column of the List of the Collection.

THE HISTORY OF THE COLLECTION.

UNTIL nearly fifty years after the establishment of the British Museum, meteorite collections nowhere existed, for the reports of the fall of stones from the sky were then treated as absurd, and the exhibition of such stones in a public museum would have been a matter for ridicule; a few stones, which had escaped destruction, were scattered about Europe, and were in the possession of private individuals curious enough to preserve bodies concerning the fall of which upon our globe such reports had been given. Hence it happened that in 1807 probably not more than four or five meteoric stones were in the British Museum; one of them was a stone of the *L'Aigle* fall, presented in 1804 by Biot, the distinguished physicist. A fragment of the *Pallas* meteorite had been presented to the Museum by the Academy of Sciences of St. Petersburg as early as 1776, at which time it was regarded as "native iron."

In the year 1807, happily for the future development of the Mineral Collection, Mr. Charles König, the mineralogist, was appointed "assistant librarian," and six years later was promoted to the Keepership of the then undivided Natural History Department; it thus came about that for thirty-eight years the senior officer of the Natural History Department of the Museum was one who had an intense enthusiasm for minerals and made them his own special study. It was in König's time (1810) that Parliament voted a special grant of £14,000 for the purchase of the minerals which had belonged to Sir Charles Greville; with these passed into the possession of the Trustees probably several fragments of meteorites, including at least one, namely *Tabor*, which had been acquired by Greville with the mineral cabinet of Baron Born. The increase of the Natural History Collections was such that in 1827 the Botanical, and in 1836 the Zoological specimens, were assigned to special departments, after which König, as Keeper of a Department thenceforward styled " Mineralogy including Geology," was left free to devote his attention to that branch of Natural History to which he was more particularly attached.

During König's time, though numerous and excellent mineral specimens were acquired, no great effort was made to render the meteorite collection itself complete ; at his death in 1851, it numbered about 68 specimens, all of them acquired by presentation or purchase ; many of the purchases were made from Mr. Heuland. The presentations were :—

One of the *Stannern* stones : by the Imperial Museum of Vienna in 1814.

Fragments of stones of the *Mooresfort* fall : by J. G. Children, Esq., F.R.S., in 1817, and by Dr. Blake in 1819.

A fragment of a stone of the *Limerick* fall : by Dr. Blake, in 1819.

The large *Tucuman* iron, and a piece of the *Imilac* siderolite : by Sir Woodbine Parish, K.C.B., F.R.S., in 1826 and 1828 respectively.

One of the *Krakhut* stones : by Wm. Marsden, Esq., in 1834.

Specimens of the *Cold Bokkeveldt* meteorite : by Sir John Herschel, F.R.S., Sir Thos. Maclear, F.R.S., and E. Charlesworth, Esq., F.G.S., in 1839.

After the death of Mr. König, Mr. C. R. Waterhouse, the palæontologist, was appointed Keeper of the Department. It was natural that the geological side of the department should then have its turn of special development, and in fact the geological collections, already important, increased from that time with great rapidity ; the mineralogical side, however, had additions made to it, though not in the proportion allotted during the preceding years. During the time of Mr. Waterhouse, only three meteorites were added to the collection, two of them by purchase ; the third, that of *Madoc*, was presented in 1856 by Sir Wm. E. Logan, F.R.S.

In the year 1857, a further division of the Collections took place, and the Minerals were placed in the Keepership of Prof. Story-Maskelyne. Under him the Mineral Collection was rendered as complete as possible in all its branches; and it is owing entirely to the unflagging energy he displayed, both in the search for, and the securing of the best obtainable specimens, that the Mineral Collection has attained to its present position of general excellence. Perhaps the greatest relative advance was made in the Collection of Meteorites. Perceiving that only half of the falls represented at Vienna were represented in the British Museum, and that the difficulty of making a fairly complete collection of such bodies must increase enormously as time went on, owing to the absorption of the specimens by public museums, Mr. Maskelyne immediately after his appointment tried to fill up the gaps. In the first place, the meteorite collections of Dr. Krantz, Mr. R. P. Greg, and Mr. R. Campbell, and many meteorites belonging to Mr. Wm. Nevill and Prof. C. U. Shepard, were acquired by purchase in 1861–2. At the same time an appeal for the donation of these bodies was sent to nearly every part of the world, and in response were presented to the Museum (1861–3) the whole or parts of many meteorites :—

From Russia.—*Tula:* by Dr. Auerbach of Moscow.

From India.—*Durala, Shalka, Bustee,* and *Dhurmsala:* by the Secretary of State for India in Council.

Moradabad, Butsura, Futtehpur, Umballa, Mhow, Manegaum, Assam, Segowlie and *Khiragurh:* by the Royal Asiatic Society of Bengal.

Nellore and *Parnallee:* by Sir W. Denison, K.C.B.

Pegu and *Kusiali:* by Dr. Thos. Oldham, F.R.S.

Kaee: by Sir Thos. Maclear, F.R.S.

Dhurmsala: by G. Lennox Conyngham, Esq.

From Australia.—The large *Cranbourne* iron: by James Bruce, Esq.

From South America.—*Vaca Muerta:* by Prof. Domeyko of Santiago.

An *Atacama* iron: by Lewis Joel, Esq.

From North America.—A specimen of the *Tucson* iron: by the Town Authorities of San Francisco.

Further, Mr. Maskelyne proposed to make the Collection more complete by exchange of fragments with other museums: and this proposition was soon accepted as peculiarly advantageous in the case of meteorites. During the same interval (1861-3), exchanges were made with the museums of Paris, Vienna, Berlin, Copenhagen, Heidelberg, and Göttingen, through Professors Daubrée, Haidinger, Rose, Hoff, Bunsen, and Wöhler, respectively: and also with the following private collectors:—Dr. Abich of Dorpat, Dr. Auerbach of Moscow, Mr. R. P. Greg of Manchester, Prof. C. U. Shepard of New Haven, U.S.A., and Dr. Sismonda of Turin.

The grand result was that by 1863, within six years of Mr. Maskelyne's appointment, the number of meteoric falls represented in the collection had been more than trebled.

Meanwhile, although Mr. Maskelyne, with the help of a single assistant (Mr. Thomas Davies), was then rearranging the general collection of minerals according to a new system of classification, time was found for a scientific examination of the meteorites thus being acquired. At that time the department was without a laboratory, and not even a blow-pipe could be used, owing to the necessity of guarding against a possible destruction of the Museum by fire. Hence recourse was had to the microscope, and as early as 1861, a microscope fitted with a graduated revolving stage and an eye-piece goniometer was constructed, under the Keeper's directions, for the examination of thin sections of meteorites with the aid of polarised light.

Working in this way, and with the simplest chemical tests, Mr. Maskelyne was the first to announce in 1862 the discovery in the Bustee meteorite of a mineral, unknown in terrestrial mineralogy, to which he gave the name of

Oldhamite, and in 1863, the more than probable occurrence
of Enstatite as an important meteoritic ingredient (Nellore).
This method of determining the mineral constituents of
a rock-section by means of the relation of the vibration-
traces to known crystallographic lines, thus first em-
ployed for the discrimination of the minerals in meteorites,
is now in general use in the investigation, not only of
meteoric, but of terrestrial rocks. About the same time,
from the Breitenbach meteorite were extracted crystals of
bronzite, which yielded the first crystallographic elements
obtained for that mineral : the measurements were made and
published by Dr. Viktor von Lang, then assistant in the
department (1862-3) and now Professor of Physics at Vienna.

The microscope was further applied to the mechanical
separation of the different mineral ingredients of a meteorite :
and by picking out in this toilsome manner the different
mineral ingredients from the crumbled material of the
Bustee aerolite, and from the residue of the Breitenbach
siderolite left after the iron had been removed by mercuric
chloride, the several silicates contained in these meteorites
were isolated for future analysis. From the particles of
colourless mineral thus obtained from the Breitenbach meteorite, one kind was selected in 1867, of which the crystals
presented a zone of orthosymmetry containing two optic
axes, and yielded two similar cleavages in a zone perpen-
dicular to the former. This ingredient was afterwards (1869)
announced to consist wholly of silica, a substance which
previous to the isolation of this mineral was only known to
occur as quartz, when in crystals, and these belong to the
hexagonal system : to the new mineral Mr. Maskelyne
later assigned the name of Asmanite. In 1868 was published by Vom Rath the discovery of a species of terrestrial
silica, the crystals of which were regarded as belonging to
the hexagonal system, though their angular elements were
distinct from those of quartz : this mineral, named by him
tridymite, has since been found (1878) to present optical and
other characters inconsistent with true hexagonal symmetry,
d is probably identical with the meteoric asmanite.

Further, another mineral occurring as minute gold-yellow

octahedra in the Bustee meteorite was recognised as new to mineralogy, and termed Osbornite.

It was not till 1867, when a laboratory was fitted up outside the Museum precincts, that it became possible to make a complete chemical examination of these materials, which had been gradually prepared and carefully picked for analysis. At Prof. Maskelyne's suggestion, the late Dr. Walter Flight was in that year appointed to assist in the laboratory-work of the Department, and gave valuable help in the chemical analysis of the above materials : the results were quite confirmatory of those already obtained by aid of the microscope and the simple tests.

Since the great increase made during the first six years of Prof. Maskelyne's Keepership, the Collection has continued to grow, though necessarily at a less rapid rate.

Of the specimens added after 1863, the following have been presented :—

1864–7 : *Manbhoom, Pokhra,* and *Muddoor :* by Dr. Thos. Oldham, F.R.S., of Calcutta.

1864 : *Atacama* (stone): by Alfred Lutschaunig.

1865–7 : *Supuhee, Jamkheir, Shytal, Udipi,* and *Lodran :* by the Government of India.

1865 : *Nerft :* by Prof. Grewingk of Dorpat.

1865 : *Ski :* by Prof. Kjerulf of Christiania.

1867–70 : *Sherghotty, Gopalpur, Khetrie, Pulsora,* and *Moteeka Nugla :* by the Trustees of the Indian Museum, Calcutta.

1867–75 : *Knyahinya* and *Zsadány :* by the Hungarian Academy of Sciences.

1869 : *Krähenberg :* by Dr. Neumayer of Pfalz.

1871 : *Searsmont :* by Dr. A. C. Hamlin of Maine, U.S.A.

1873 : *Stannern* and *Great Fish River :* by Dr. Benj. Bright of Bristol.

1874 : *Great Namaqualand :* by the South African Museum.

1875 : *West Liberty :* by Dr. G. Hinrichs of Iowa, U.S.A.

1876 : *Shingle Springs :* by E. N. Winslow, Esq.

1876 : *Rowton :* by the Duke of Cleveland.

1877: *Khairpur* and *Jhung:* by A. Brandreth, Esq., of Calcutta.

1877: *Verkhne-Dnieprovsk:* by Prof. Koulibini of St. Petersburg.

1878 : *Cronstadt :* by John Sanderson, Esq., of Natal.

1878: *Santa Catharina:* by Prof. Daubrée of Paris.

1879: *Imilac, Serrania de Varas,* and *Mount Hicks:* by George Hicks, Esq., of Newquay.

1881 : *Middlesbrough:* by the Board of Directors of the North Eastern Railway.

1882 : *Veramin :* by the Shah of Persia.

1882 : *Vaca Muerta:* by F. A. Eck, Esq., of London.

1883: *Ogi:* by Naotaro Nabeshima, Esq., formerly Daimiô of Ogi, Japan.

1885: *Ivanpah:* by H. G. Hanks, Esq., of San Francisco.

1885 : *Youndegin :* by Rev. Charles G. Nicolay of Western Australia.

1885 *et seq.*: *Chandpur, Pirthalla, Nammianthal, Lalitpur, Heidelberg, Wöhler's iron, Wessely, Nageria, Esnandes, Kahangarai, Bori, Bishunpur* and *Ambapur Nagla:* by the Director of the Geological Survey of India.

1885: *Lucky-Hill:* by the Governors of the Jamaica Institute.

1886: *Nenntmannsdorf:* by Dr. H. B. Geinitz of Dresden.

1886: *Jenny's Creek:* by John N. Tilden, Esq., of New York State, U.S.A.

1887 : *Djati-Pengilon:* by the Government of the Netherlands.

1887 : *Glorieta Mountain:* by Richard Pearce, Esq., of Colorado, U.S.A.

1889 : *Kalambi* and *Bhagur:* by the Bombay Branch of the Royal Asiatic Society.

1890 : *Bendegó River:* by the Director of the National Museum, Rio de Janeiro.

1891: *Dundrum:* by the Board of Trinity College, Dublin.

1891: *Washington :* by G. F. Kunz, Esq., of New York, U.S.A.

1891: *Thunda :* by Prof. A. Liversidge, F.R.S., of Sydney.

1894: *Makariwa :* by Prof. G. H. F. Ulrich, F.G.S., of Dunedin, New Zealand.

1894: *Bherai :* by His Highness the Nawab of Junagadh, India.

1896: *Madrid :* by Don Miguel Menino of Madrid.

Since the same year (1863) exchanges have been made with the museums of Belgrade, Berlin, Blömfontein, Breslau, Calcutta, Cambridge, Christiania, Debreczin, Dresden, Fremantle, Göttingen, Odessa, Paris, Pau, Rio de Janeiro, Rome, South Africa, Stockholm, Transylvania, Troyes, Utrecht, Vienna, Washington, and Yale College; and also with the following :—Dr. Abich of Dorpat, Dr. Auerbach of Moscow, S. C. H. Bailey, Esq., of Cortlandt-on-Hudson, U.S.A., Prof. Baumhauer of Haarlem, C. S. Bement, Esq., of Philadelphia, U.S.A., Dr. Breithaupt of Freiberg, J. R. Gregory, Esq., of London, Prof. C. T. Jackson of Boston, U.S.A., Henry Ludlam, Esq., of London, Prof. W. Mallet of Virginia, U.S.A., Prof. Vom Rath of Bonn, Prof. C. U. Shepard of New Haven, U.S.A., His Excellency Julien de Siemachko of St. Petersburg, Prof. Lawrence Smith· of Louisville, U.S.A., and J. N. Tilden, Esq., of New York.

In this way, by the generosity and self-denial of donors, by the somewhat difficult method of exchange, and by purchase, it has been possible to get together the fine representative collection of meteorites now in the British Museum.

AN INTRODUCTION

STUDY OF METEORITES.

Most of the specimens here referred to are in Case 4 in the Pavilion at the end of the Mineral Gallery.

The fall of stones from the sky formerly discredited.   1. Till the beginning of the present century, the fall of stones from the sky was an event, the actuality of which neither men of science nor the mass of the people could be brought to believe in. Yet such falls have been recorded from the earliest times, and the records have occasionally been received as authentic by a whole nation. In general, however, the witnesses of such an event have been treated with the disrespect usually shown to reporters of the extra-ordinary, and have been laughed at for their supposed delusions : this is less to be wondered at when we remember that the witnesses of a fall have usually been few in number, unaccustomed to exact observation, frightened by what they both saw and heard, and have had a common tendency towards exaggeration and superstition.

Ancient records.   2. The most ancient of all such records, if interpreted in the usual way, is that given in the tenth chapter of the Book of Joshua, where we read that during the flight of the Canaanites after the battle of Gibeon, great stones were cast down from heaven, so that more were slain by them than with the sword. It is not quite clear, however, from the text that a prolonged shower of large hailstones is not referred to.

B

A stone, famous through long ages,[*] fell in Phrygia and was preserved there for many generations. About 204 B.C. it was demanded from King Attalus and taken with great ceremony to Rome. It is described as "a black stone, in the figure of a cone, circular below and ending in an apex above." In his History of Rome, Livy tells of a shower of stones on the Alban Mount, about 652 B.C., which so impressed the senate that a nine days' solemn festival was decreed. Other instances of the "rain of stones" in Italy are mentioned by the same author. Plutarch relates the fall of a stone in Thrace about 470 B.C., during the time of Pindar, and according to Pliny, the stone was still preserved in his day, 500 years afterwards. The latter records two other falls, one in Asia Minor, the other in Macedonia.

De Guignes in his Travels states that, according to old Chinese manuscripts, falls of stones have again and again been observed in China; the earliest mentioned is one which happened about 644 B.C.

Worship of meteoric stones.

3. These falls from the sky, when credited at all, have been deemed prodigies or miracles, and the stones have been regarded as objects for reverence and worship. It has even been conjectured that the worship of such stones was the earliest form of idolatry. The Phrygian stone, mentioned above, was worshipped at Pessinus by the Phrygians and Phœnicians as Cybele, "the mother of the gods," and its transference to Rome followed the announcement by an oracle that possession of the stone would secure to the state a continual increase of prosperity. Similarly, the Diana of the Ephesians, "which fell down from Jupiter," and the image of Venus at Cyprus appear to have been, not statues, but conical or pyramidal stones. A stone, of which the history goes back far beyond the seventh century, is still revered by the Moslems as one of their holiest relics, and is preserved at Mecca built into the north-eastern corner of the Kaaba. The late Paul Partsch,[†] for

[*] Remarks concerning stones said to have fallen from the clouds both in these days and in ancient times : by Edward King. London, 1796. Mémoire historique et physique sur les chutes des pierres : par P. M. S. Bigot de Morogues. Orléans, 1812.

[†] Sitzungsber. d. k. Ak. d. Wiss. Wien. 1856, vol. 22, p. 393.

many years Keeper of Minerals in the Imperial Museum of Vienna, considered that the meteoric origin of the Kaaba stone was sufficiently proved by descriptions which had been submitted to him. A stone which fell in Japan about the year 1730, and was lately presented to the British Museum, was long made an annual offering in a temple of Ogi at one of the Japanese religious festivals. It may be added that a stone which lately fell in India* was decked with flowers, daily anointed with ghee (clarified butter), and subjected to frequent ceremonial worship and coatings of sandal-wood powder. The stone was placed on a terrace constructed for it at the place where it struck the ground, and a subscription was made for the erection of a shrine.

The oldest undoubted meteoric stone still preserved. 4. The oldest undoubted sky-stone still preserved is that which, though after the Revolution removed for a time to the Library at Colmar, is once more suspended by a chain from the vault of the choir of the parish church of Ensisheim in Elsass. The following is a translated extract from a document kept in the church :—

"On the 16th of November, 1492, a singular miracle happened : for between 11 and 12 in the forenoon, with a loud crash of thunder and a prolonged noise heard afar off, there fell in the town of Ensisheim a stone weighing 260 pounds. It was seen by a child to strike the ground in a field near the canton called Gisgaud, where it made a hole more than five feet deep. It was taken to the church as being a miraculous object. The noise was heard so distinctly at Lucerne, Villing, and many other places, that in each of them it was thought that some houses had fallen. King Maximilian, who was then at Ensisheim, had the stone carried to the castle : after breaking off two pieces, one for the Duke Sigismund of Austria and the other for himself, he forbade further damage, and ordered the stone to be suspended in the parish church."

* Records of the Geological Survey of India. Calcutta, 1885, vol. 18, p. 237.

Scientific
men begin
to investi-
gate the
reports.   5. Three French Academicians, one of whom was the afterwards renowned chemist Lavoisier, presented to the Academy in 1772 a report on the analysis of a stone said to have been seen to fall at Lucé on September 13, 1768. As Pane 4c. the identity of lightning with the electric spark had been recently established by Franklin, they were in advance convinced that "thunder-stones" existed only in the imagination; and never dreaming of the existence of a "sky-stone" which had no relation to a "thunder-stone," they somewhat easily assured both themselves and the Academy that there was nothing unusual in the mineralogical characters of the Lucé specimen, their verdict being that the stone was an ordinary one which had been struck by lightning.

Chladni
argues that
the bodies
come from
outer space.   6. In 1794 the German philosopher Chladni, famed for his researches into the laws of sound, brought together numerous accounts of the fall of bodies from the sky, and called the attention of the scientific world to the fact that several masses of iron, of which he specially considers two, had in all probability come from outer space to this planet.*

The Pallas
iron.   One of them is the mass still known as the Pallas or Pane 4c. Krasnojarsk iron.† This irregular mass, weighing about 1500 lbs., of which the greater part is still in the Museum at St. Petersburg, was met with at Krasnojarsk by the traveller Pallas in the year 1772, and had been found in 1749 by a Cossack on the surface of the highest part of a lofty mountain between Krasnojarsk and Abakansk in Siberia, in the midst of a schistose district: it was regarded by the Tartars as a "holy thing fallen from heaven." The interior is composed of a ductile iron, which, though brittle at a high temperature, can be forged either cold or at a moderate heat; its large sponge-like pores are filled with an amber-coloured olivine; the texture is uniform, and the olivine equally distributed; a vitreous varnish had preserved it from rust. The fragment in the case, weighing about

---

* Ueber den Ursprung der von Pallas gefundenen und anderer ihr ähnlicher Eisenmassen. Riga, 1794.

† Reise durch verschiedene Provinzen des russischen Reichs: von P. S. Pallas. St. Petersburg, 1776, Part III., p. 411.

7 lbs., was presented in 1776 by the Imperial Academy of Sciences of St. Petersburg.

The Otumpa iron. A second specimen referred to is that which in 1783 Don Michael Rubin de Celis was sent by the Viceroy of Rio de la Plata to investigate; * it had been found by Indians, searching for honey and wax, and trusting to rain for drink, projecting about a foot above the ground near a place called Otumpa, in the Gran Chaco Gualamba, South America, and was at first thought to be the outcrop of an iron vein. Don Rubin de Celis estimated the weight of this mass of malleable iron at thirty thousand pounds, and reported that for a hundred leagues around there were neither iron mines nor mountains nor even the smallest stones, and that owing to the absence of water, there was not a single fixed habitation in the country. There were several smaller masses at the locality ; one of them, weighing 1400 lbs., is shown on a separate stand in the Pavilion: according to Sir Woodbine Parish, who presented it to the Museum in 1826, it had been removed to Buenos Ayres at the beginning of the struggle for Independence ; it was a complimentary gift to Sir Woodbine on the occasion of his being sent by Canning to acknowledge the Independence of the State. A slice of this iron is shown in case 4c. **Pane 4c.**

Separate stand.

Chladni's arguments. 7. Chladni argued that these masses could not have been formed in the wet way, for they had evidently been exposed to fire and slowly cooled: that the absence of scoriæ in the neighbourhood, the extremely hard and pitted crust, the ductility of the iron, and, in the case of the Siberian mass, the regular distribution of the pores and olivine, precluded the idea that they could have been formed where found, whether by man, electricity, or an accidental conflagration : he was driven to conclude that they had been formed elsewhere, and projected thence to the places where they were discovered; and as no volcanoes had been known to eject masses of iron, and as, moreover, no volcanoes are met with in those regions, he held that the specimens referred to must have actually fallen from the sky. Further, he sought to show that the flight of a heavy body through the sky is the direct cause of the luminous phenomenon known as a fire-ball.

* Philosophical Transactions. London, 1788, vol. 78, part 1, pp. 37, 183.

The fall of
stones at
Siena, in
Tuscany.

8. About seven o'clock on the evening of June 16, 1794, Pane 4a. as if to direct attention to Chladni's theory, there fell a shower of stones at Siena, in Tuscany.

The event is described in the following letter, dated Siena, July 12, 1794, from the Earl of Bristol to Sir William Hamilton, K.B., F.R.S., at that time British Envoy-Extraordinary and Plenipotentiary at the Court of Naples:—*

"In the midst of a most violent thunderstorm, about a dozen stones of various weights and dimensions fell at the feet of different persons, men, women and children. The stones are of a quality not found in any part of the Siennese territory; they fell about 18 hours after the enormous eruption of Mount Vesuvius: which circumstance leaves a choice of difficulties in the solution of this extraordinary phenomenon. Either these stones have been generated in this igneous mass of clouds which produced such unusual thunder, or, which is equally incredible, they were thrown from Vesuvius, at a distance of at least 250 miles: judge, then, of its parabola. The philosophers here incline to the first solution. I wish much, Sir, to know your sentiments. My first objection was to the fact itself, but of this there are so many eyewitnesses, it seems impossible to withstand their evidence."

The fall of
a stone
near Wold
Cottage,
Yorkshire.

9. Soon afterwards there fell a stone in England itself. Pane 4b. About three o'clock in the afternoon of December 13, 1795, a labourer working near Wold Cottage, a few miles from Scarborough, in Yorkshire,† was terrified to see a stone fall about ten yards from where he was standing. The stone, weighing 56 lbs., was found to have gone through 12 inches of soil and 6 inches of solid chalk rock. No thunder, lightning, or luminous meteor accompanied the fall; but in the adjacent villages there was heard an explosion likened by the inhabitants to the firing of guns at sea, while in two of them the sounds were so distinct of something singular

---

* Philosophical Transactions. London, 1795, vol. 85, p. 103.
† *Ibid.*, 1802, vol. 92, p. 174.

passing through the air towards Wold Cottage, that five or six people went to see if anything extraordinary had happened to the house or grounds. No stone presenting the same characters was known in the country. The stone is preserved in the Museum Collection.

Terrestrial origin still sought for.

10. It seemed to be now impossible for any one to doubt the fall of stones from the sky, but the reluctance of scientific men to grant an extra-terrestrial origin to them is shown by the theories referred to in the above letter to Sir William Hamilton, and is rendered even more evident by the theory proposed ·in 1796 by Edward King, who suggested that the stones had their origin in the condensation of a cloud of ashes, mixed with pyritical dust and numerous particles of iron, coming from some volcano. As the stones fell at Siena out of a cloud coming from the North, while Vesuvius is really to the South, he gravely suggested that in this case the cloud had been blown from the South past Siena, and had then before its condensation been brought back by a change of wind. As to the fall of a stone near Wold Cottage, he was not prepared either to believe or disbelieve the witnesses until the matter had been more closely examined; but in case the statements should prove worthy of credit, he points out the possibility of the necessary cloud having come from Mount Hecla in Iceland.

The fall of stones near Benares, in India.

11. Later came a well-authenticated account of a more wonderful event still. At 8 o'clock on the evening of December 19, 1798, many stones fell at Krakhut, 14 miles from Benares, in India; the sky was perfectly serene, not a cloud had been seen since December 11th, and none was seen for many days after. According to the observations of several Europeans, as well as natives, in different parts of the country, the fall of the stones was preceded by the appearance of a *ball of fire*, which lasted for only a few instants, and was followed by an explosion resembling thunder. Pane 4o

Examination of stones by Howard.

12. Fragments of the stones of Siena, Wold Cottage, and Krakhut, as also of a stone said to have fallen on July 3, 1753, at Tabor, in Bohemia, came into the hands of Edward Howard, and the comparative results of a chemical and mineralogical investigation (the latter by the Count de

Bournon) of the stones from the above four places are given in a paper read before the Royal Society of London, on February 25, 1802. Howard concludes as follows:—

"The mineralogical descriptions of the Lucé stone by the French Academicians, of the Ensisheim stone by M. Barthold, and of stones from the above four places (Siena, Wold Cottage, Krakhut and Tabor) by the Count de Bournon, all exhibit a striking conformity of character common to each of them, and I doubt not but the similarity of component parts, especially of the malleable alloy, together with the near approach of the constituent proportions of the earth contained in each of the four stones, will establish very strong evidence in favour of the assertion that they have fallen on our globe. They have been found at places very remote from each other, and at periods also sufficiently distant. The mineralogists who have examined them agree that they have no resemblance to mineral substances properly so called, nor have they been described by mineralogical authors."

Pane 4c.

Could projectiles reach the earth from the moon?

13. This paper aroused much interest in the scientific world, and, though Chladni's theory that such stones come from outer space was still not generally accepted in France, it was there deemed more worthy of consideration after Poisson* (following Laplace) had shown that a body shot from the moon in the direction of the earth, with an initial velocity of 7592 feet a second, would not fall back upon the moon, but would actually, after a journey of sixty-four hours, reach the earth, upon which, neglecting the resistance of the air, it would fall with a velocity of about 31,508 feet a second.

The fall of stones at L'Aigle, in France.

14. Whilst the minds of the scientific men of France were in this unsettled condition, there came a report that still another shower of stones had fallen, this time in their own country, and within easy reach of Paris. To settle the matter finally, if possible, the physicist Biot, Member of the French Academy, was directed by the Minister of the Interior to

Pane 4c.

* Bulletin des Sciences par la Société Philomathique. Paris, 1803, vol. 3, no. 71, p. 180.

inquire into the event upon the spot. After a careful examination of the stones and a comparison of the statements of the villagers, Biot* was convinced that—

1. On Tuesday, April 26, 1803, about 1 P.M., there was a violent *explosion* in the neighbourhood of L'Aigle, in the department of Orne, lasting for five or six minutes: this was heard for a distance of 75 miles round.

2. Some moments before the explosion at L'Aigle, a *fireball* in quick motion was seen from several of the adjoining towns, though not from L'Aigle itself.

3. There was absolutely no doubt that on the same day many stones fell in the neighbourhood of L'Aigle.

Biot estimated the number of the stones at two or three thousand; they fell within an ellipse of which the larger axis was 6·2 miles, and the smaller 2·5 miles; and this inequality might indicate not a single explosion but a series of them. With the exception of a few little clouds of ordinary character, the sky was quite clear.

The exhaustive report of Biot, and the conclusive nature of his proofs, compelled the whole of the scientific world to recognise the fall of stones on the earth from outer space as an undoubted fact.

The times and places of fall are independent of terrestrial circumstances.

15. Since that date many falls have been observed, and the attendant phenomena carefully investigated. These observations teach us that *meteorites*, as they are now called, fall at all times of the day and night, and at all seasons of the year, while they favour no particular latitudes: also they are found to be quite independent of the weather, and in many cases have fallen when the sky has been perfectly clear; even where stones have fallen in what has been called a thunder-storm, we may reasonably suppose that in most cases the luminous phenomena have been mistaken for lightning, and the noise of the explosion for thunder.

Velocity of meteorites.

16. From observations of the path and the time of flight, it is calculated that meteorites enter the atmosphere with

* Mémoires de l'Institut National de France. 1806, vol. 7, part 1, Histoire, p. 224.

velocities ranging from 10 to 45 miles a second. Let us attempt to follow the course of a body moving at such a rate. So long as the body is traversing " empty space," the only heat it receives is that sent direct from the sun ; the meteorite will thus be probably very cold, and, from its small size and want of luminosity, invisible to an observer on the earth's surface. After the meteorite enters the earth's atmosphere a very speedy change must take place.

The resistance of the air.
Assuming the law of resistance of the air for a planetary velocity to be the same as that deduced from experiments with artillery, the astronomer Schiaparelli[*] has shown that if a ball of 8 inches diameter and $32\frac{1}{3}$ lbs. weight enter the atmosphere with a velocity of $44\frac{3}{4}$ miles a second, its velocity on arriving at a point where the barometric pressure is still only $\frac{1}{760}$-th of that at the earth's surface will have been already reduced to $3\frac{1}{2}$ miles a second. From this it is clear that the speed of the meteorite after the whole of the atmosphere has been traversed will be extremely small, and comparable with that of an ordinary falling body. From experiments lately made by Professor A. S. Herschel, it has been calculated that the velocity of the meteorite which fell at Middlesbrough, in Yorkshire, on March 14, 1881, was, on striking the ground, only 412 feet a second. In the case of the Hessle fall, several stones fell on the ice, which was only a few inches thick, and rebounded without either breaking the ice or being broken themselves.

Transformation of the energy.
17. Further, Schiaparelli points out that in the case supposed, the energy already converted into heat would be sufficient to raise 198,400 pounds of water from freezing point to boiling point under the ordinary barometric pressure. The greater part of this heat is, no doubt, carried off by the air through which the meteorite passes ; but still the wonder is, not that a meteorite is small on reaching the earth's surface, but that any of it is left to " tell the tale."

The cloud, ball of fire and train.
This sudden generation of heat will cause a fusion and volatilisation of the outer material of the meteorite, and in

[*] Principes de Thermodynamique : par Paul de Saint-Robert. Paris, 1870, p. 329.

some cases a combustion of some of its constituents : the products of this action sufficiently account for the *cloud* from which the meteorite is generally seen to emerge as a ball of fire, and also for the train often left behind. The ball of fire has often an apparent diameter larger even than that of the moon, and is sometimes too bright for the eye to gaze upon.

The meteorite is only luminous in the first part of its flight through the air. Owing to the quick reduction of speed, the luminosity will be a feature of the higher part of the course. The Orgueil meteorite of May 14, 1864, was so high when luminous that, notwithstanding its almost easterly motion, it was seen over a space of country ranging from the Pyrenees to the north of Paris, a distance of more than 300 miles.

The time of flight through the air is very brief. 18. Next we may remark that the time of flight in the earth's atmosphere will be very short, and reckoned only by seconds. Even when the meteorite is wholly metallic, if we may judge from the time one end of a poker may be held in the hand whilst the other end is in the fire, the heat will not have had time to get far below the surface before the body Pane 4d. will have reached the ground.

The crust. 19. As a matter of fact, meteorites are invariably found to be covered with a *crust* or varnish, the thinness of which shows the slight depth to which the heat has had time to penetrate ; in the case of the stones, the greater part of the suddenly heated superficial material must chip off and be left behind. The appearance of the crust varies according to the mineral constitution of the meteorites : it is generally black, and in most cases dull as in High Possil, Zsadány and Pane 4d. Orgueil, but sometimes shiny, as in Stannern, or partly dull and partly shiny, as in Dyalpur ; or it is of a dark grey colour, as in Mezö-Madaras and some of the stones which fell in the neighbourhood of Mocs. In the case of the Pultusk meteorite of January 30, 1868, several thousands Panes 4e/g. of stones, varying from the size of a man's head to that of a small nut, were picked up, each covered with a crust : fifty-seven of the stones of this fall are shown in the case.

Its ridges and furrows. 20. The crust is not of equal thickness over the whole of the meteorite, but, owing to the motion through the air, is generally in *ridges* and *furrows*, of which the directions indicate the position of the meteorite in regard to its line of motion at a certain part of its course; and this relation is rendered more clear in some cases by the position of the *swellings* produced by the flow of melted material to the back of the moving mass. The Nedagolla iron and the Goalpara stone illustrate this peculiarity. Meunier grants that the crust is due to the action of heat, but considers that the action is direct, and not through fusion: he holds that only the outer surface of the crust itself has been melted, and that the furrows and swellings are due to the scooping action of the air through which the meteorite at first passes with so enormous a velocity. Pane 4*h*.

The pittings.

21. Further, the surface of a meteorite is generally covered with *pittings*, which have been compared in form to thumb-marks: stones from the Supuhee, Futtehpur, and Knyahinya falls present good examples of this character. It is remarkable that pittings bearing a close resemblance to those of meteorites have been observed on the large partially burned grains of gunpowder, which have been picked up near the muzzle after the firing of the 35-ton and 80-ton guns at Woolwich. The pitting of the gunpowder grains is attributed to unequal combustion, but that of meteorites seems to be due not so much to inequality of combustibility as to that of conductivity and fusibility of the matter at the surface. Pane 4*h*. Pane 4*h*.

Fragmentary form of meteorites.

22. As picked up, complete and covered with crust, meteorites are not spherical, nor have they any definite shape: in fact, they are always irregular angular fragments, such as would be obtained on breaking up a rock presenting no regularity of structure.

The explosions.

23. The sudden generation of heat, and the consequent expansion of the outer shell, account not only for the *break-up* of the meteorite into fragments, but partly also for the *crash like that of thunder* which is a usual accompaniment of the fall. Some refer this noise solely to the sudden rush of air into the vacuum which is so quickly left behind by the meteorite in the early part of the course. In the consideration of this question the Butsura fall of May 12, 1861, is

particularly interesting.* The detonations, in this case three in number, were heard 60 miles away at Goruckpur. Fragments of the stone were picked up three or four miles apart, and, wonderful to say, it was possible to reconstruct Pane 4*λ*. with much certainty the portion of the meteorite of which they are the part: a model of the reconstructed portion is shown in the case. Two of the fragments, in other respects Pane 4*a*. fitting perfectly together, are even on the faces of the junction now coated with a black crust, showing that one disruption took place when the meteorite had a high velocity; two other fragments found some miles apart fitted perfectly, and were neither of them incrusted at the surface of fracture, thus indicating another disruption at a time when the velocity of the meteorite had been so far reduced that the material of the new faces was not melted through the generation of heat. Sometimes, as in the case of the meteorite of Orgueil, the fragments reach the ground before the detonation is heard, proving that the fracture has taken place at a part of the course where the velocity of the meteorite was considerably greater than that of the sound-vibrations (1100 feet a second).

The sounds heard after the loud explosions. 24. After the detonation are generally heard sounds which have been variously likened to the flapping of the wings of wild geese, the bellowing of oxen, Turkish music, the roaring of a fire in a chimney, the noise of a carriage on the pavement, and the tearing of calico: these sounds are probably due to the whirling of the fragments through the air in the neighbourhood of the observers.

The chemical elements found in meteorites. 25. As to the *kinds of elementary matter*† of which meteorites are composed, about one-third, and those the most common, of the elements at present recognised as constituents of the earth's crust have been met with: no new elementary body has been discovered.

* The Fall of Butsura: by Prof. Maskelyne. Phil. Mag. 1863, vol. 25, p. 50.

† Die chemische Natur der Meteoriten: von C. Rammelsberg. Berlin, 1870-9.

Météorites: par S. Meunier. Paris, 1884.

The most frequent or plentiful in their occurrence are :—

Iron	Oxygen
Nickel	Silicon
Phosphorus	Magnesium
Sulphur	Calcium
Carbon	Aluminium :

while, less frequently or in smaller quantities, are found :

Hydrogen	Chromium
Nitrogen	Manganese
Chlorine	Cobalt
Lithium	Arsenic
Sodium	Antimony
Potassium	Tin
Strontium	Copper.
Titanium	

Elements of doubtful presence. 26. In addition to the above the existence of traces of several other elements has been announced, but the accuracy of their determination is not beyond doubt: lead is undoubtedly present in the Tarapaca iron, but was probably artificially introduced.

Both simple and combined. 27. All the elements are present in the combined state ; the iron occurring chiefly as an alloy with nickel, and the phosphorus almost always combined with both nickel and iron. Some of them are found also in their elementary condition ; hydrogen and nitrogen, as occluded gases, and carbon both as indistinctly crystallised diamond and as graphitic carbon, the latter being generally amorphous, but occasionally in cubic crystals as cliftonite ; free sulphur has been observed in one of the carbonaceous meteorites, but may have been separated from the unstable sulphides since the entry into our atmosphere.

Some of the constituents are new to mineralogy. 28. Of the constituents of meteorites, the following are by many mineralogists regarded as being at present unrepresented among the terrestrial minerals :— Pane 4k.

Cliftonite, a cubic form of graphitic carbon,
Various alloys of nickel and iron,

Schreibersite, phosphide of iron and nickel,
Troilite, proto-sulphide of iron,
Oldhamite, sulphide of calcium,
Osbornite, oxy-sulphide of calcium and titanium or
   zirconium,
Daubréelite, sulphide of iron and chromium,
Lawrencite, protochloride of iron,
Cohenite, a carbide of iron and nickel,
Asmanite, a species of silica,
Maskelynite, a singly refracting mineral with the com-
   position of labradorite.

Nature of troilite and asmanite. Of the above, *troilite* is perhaps identical with some varieties of terrestrial pyrrhotite: *asmanite*, the form of silica obtained in 1867 by Maskelyne from the Breitenbach meteorite, was announced by him in 1869 to be optically biaxal, and thus to belong to a crystalline system different from the hexagonal to which both tridymite, then just announced by Vom Rath, and quartz had been assigned. Later investigations of tridymite have shown that its optical characters and crystalline form are inconsistent with the hexagonal system of crystallisation, and it is not impossible that asmanite and tridymite may be identical. It has been found that tridymite becomes optically uniaxal at a moderate temperature, and its general characters appear to be essentially identical with those of asmanite.

Compounds identical with terrestrial minerals. 29. Other compounds are present, corresponding to the Pane 44. following terrestrial minerals:—

   Olivine,
   Enstatite and bronzite,
   Diopside and augite,
   Anorthite and labradorite,
   Magnetite and chromite,
   Pyrites,
   Pyrrhotite,
   Breunnerite.

Further, from one of the Lancé stones, chloride of sodium and from the carbonaceous meteorites, sulphates of sodium, calcium, and magnesium have been extracted by means of water; carbonic oxide, carbonic acid and marsh gas have been found as occluded gases.

In addition to the above, there are several compounds or mixtures of which the nature is not satisfactorily ascertained.

The rarity of quartz. **30.** Quartz, the most common of terrestrial minerals, is absent from the stony meteorites; but in the undissolved residue of the Toluca iron microscopic crystals have been found, some of which have important characters identical with those of quartz, while others resemble zircon. As mentioned above, free silica is found in the Breitenbach meteorite as asmanite.

The conditions under which these compounds can have been formed. **31.** As to the *conditions** under which such compounds can have been formed, we may assert that they must have been very different from those which at present obtain near the earth's surface: in fact, it is difficult to imagine that the metallic nickel-iron and the unstable sulphides can either have been formed, or have remained undecomposed, under circumstances in which water and atmospheric air have played any prominent part. Still, what little we do know of the inner part of our globe does not shut out the possibility of the existence of similar compound and elementary bodies at great depths below the surface. Daubrée,† after experiment, inclines to the belief that the iron is due, in many cases at least, to reduction from an olivine rich in diferrous silicates, and this view acquires some additional probability from the presence of the gases hydrogen and carbonic oxide in several meteoric irons: the existence, however, of such siderolites as that of Krasnojarsk, which is rich both in metallic iron and in silicate of iron and magnesium (olivine), and yet presents no traces of the intermediate silicate of magnesium (enstatite), offers a weighty objection to the general application of this view.

Classification. **32.** Meteorites may be conveniently arranged in three classes, which pass more or less gradually into each other: the first includes all those which consist mainly of iron, and have, therefore, been called by Maskelyne aero-siderites

* Some lecture-notes on meteorites: by Prof. Maskelyne. *Nature*, 1875, vol. 12, pp. 485, 504, 520.
† Etudes synthétiques de géologie expérimentale. Paris, 1879, p. 517.

(*aer*, air, and *sideros*, iron), or more shortly, *Siderites;* the second is formed by those which are composed chiefly of iron and stone, both in large proportion, and are called aerosiderolites, or, shortly, *Siderolites;* while those of the third class, being almost wholly of stone, are called *Aerolites* (*aer*, air, and *lithos*, stone).

The siderites.

33. In the Siderites the iron generally varies from 80 to 95 per cent., and the nickel from 6 to 10 per cent.; in the Santa Catharina siderite (of which the meteoric origin is somewhat doubtful) 34, and in that of Oktibbeha County 60 per cent. of nickel have been found : the nickel is alloyed with the iron, and several of the alloys have been distinguished by special names. Owing to the presence of the nickel, meteoric iron is often so white on a fractured surface as to be mistaken for silver by its finder; it is also less liable to rust than ordinary iron is. Troilite is frequently present in veins or large nodules, sometimes surrounded by graphite; schreibersite is almost always found, and occasionally also daubréelite.

Occluded gases.

Further, the researches of various chemists have proved the presence of the gases hydrogen, nitrogen, marsh gas, and the carbonic oxides, occluded in the iron; Dr. Walter Flight has shown that the gases occluded in the Rowton iron would, under normal temperature and pressure, have a volume upwards of six times that of the meteorite itself.

Figures produced by action of acids.

34. The want of homogeneity and the structure of meteoric iron are beautifully shown by the figures generally called into existence when a polished surface is exposed to the action of acids or bromine; they are due to the inequality of the action on the various constituents, and the layers are composed chiefly of kamacite and of tænite, alloys of nickel and iron. In the Agram iron, investigated by Widmanstätten in 1808, the layers are parallel to the faces of the regular octahedron; such figures are well shown by the exhibited slice of the Toluca iron; different degrees of distinctness of such " Widmanstätten" figures are illustrated by specimens of Seneca River, Zacatecas, Charcas, Burlington, Jewell Hill, Lagrange, Victoria West, Nelson County, and

Pane 4*l.*

Pane 4*l.*

See-Läsgen. The Braunau iron has cleavages parallel to the faces of a cube, and on etching yields linear furrows which were found (1848) by Neumann to have directions such as would result from twinning about an octahedral face; as illustrations of the "Neumann" figures, etched specimens of Braunau and Salt River are exhibited. The large Tucuman specimen, mounted on a separate pedestal, furnishes a good example of the less distinct, and more or less damascene, appearance presented by the etched surface of some meteoric irons.

Pane 4*l.*

Few siderites have been seen to fall.

35. The Siderites *actually observed to fall* reach only the small number of eight; they are, Agram, Charlotte, Braunau, Victoria West, Nedagolla, Rowton, Mazapil and Cabin Creek. The remaining specimens in collections of Siderites are presumed to be of meteoric origin, as suggested in Art. 7, by reason of the peculiarity of their appearance and chemical composition, and of the locality in which they have been found.

The iron found at Ovifak is probably of terrestrial origin.

36. The difficulty of distinguishing an iron of terrestrial from one of meteoric origin has been lately rendered very evident by the controversy as to the origin of the large masses of iron, containing one or two per cent. of nickel, and weighing 9,000, 20,000, and 50,000 lbs. respectively, found in 1870 by Baron Nordenskiöld on the beach at Ovifak, Disko Island, Western Greenland.

A careful examination of the rocks of the neighbourhood shows that the basalt contains nickeliferous iron disseminated through it, and that the large masses of iron, at first thought to be meteorites, are very probably of terrestrial origin, and have been left exposed upon the seashore, through the weathering of the rock which originally enclosed them. Part of a mass extracted from the rock by Professor Nordenskiöld, and presented by him to the Museum, is shown on a table in the Pavilion. Malleable metallic nodules extracted from the rock itself were found to contain as much as 6·5 per cent. of nickel. In 1880 Steenstrup[*] found ferriferous basalt *in situ* in three different parts of the island. At Assuk (Asuk) the enclosed balls of iron reach a

Pane 4*m.*

* Mineralogical Magazine. London, 1884, vol. 6, p. 1.

diameter of nearly three quarters of an inch. Some assert that the basalt and the nickel-iron have been expelled together from great depths below the earth's surface, while others consider that the nickel-iron is due to the reduction of the basalt by its passage through the beds of lignite and other vegetable matter found in the vicinity.

Other terrestrial irons.   37. With the Ovifak iron in the case are shown other specimens of iron which have been brought by various explorers from Western Greenland, and were formerly thought to have had a meteoric origin. The discovery of ferriferous basalt, not only *in situ* in several places, but also deposited in a Greenlander's grave (1879) along with knives (similar to those brought home by Ross) and the usual stone tools, renders it clear that the Esquimaux were not dependent on meteorites for their metallic iron, as had long been supposed.

Page 4m.

Mr. Skey announced in 1885 the discovery of terrestrial nickel-iron in New Zealand. Grains of the alloy (Awaruite), containing as much as 67·6 per cent. of nickel, are found in the sand of the rivers flowing from a range of mountains composed of olivine-enstatite rocks, in places altered to serpentine : similar particles have been found in the serpentine itself. Similarly, in the sand of the stream Elvo, near Biella, in Piedmont, grains of nickel-iron containing 75 per cent. of nickel have been found : and in the placer gravel of a stream in Josephine and Jackson Counties, Oregon, U.S.A., large quantities of waterworn pebbles, which enclose an alloy (Josephinite) of nickel and iron containing 72 per cent. of the former metal, have been met with. Professor Andrews many years ago established the presence of minute particles of metallic iron in some basalts ; Dr. Sauer has lately found a single nodule of malleable iron of the size of a walnut in the basalt of Ascherhübel, in Saxony, and Dr. Johnston-Lavis has announced the find of an enclosure of metallic iron in a leucitic lava of Monte Somma ; Dr. Hoffmann has noted the occurrence of minute spherules of brittle iron both in perthite and quartzite in Ontario.

The stony matter of meteorites.   38. The stony part of the siderolites and aerolites is almost entirely crystalline, and in most cases presents a peculiar " chondritic " or granular structure, the loosely coherent grains being composed of minerals similar to those which enclose them, and containing in most cases minute particles of iron and troilite disseminated through them : glass-inclusions are found to be present. The minerals mentioned above as occurring in meteorites are such as are very characteristic of the more basic terrestrial rocks, such as dunite, lherzolite and basalt, which have been expelled from considerable depths below the earth's surface.

39. Several attempts to classify aerolites according to their mineralogical constitution have been made, but it cannot be said that any of them is very satisfactory; seeing that even in the same stone there may be much difference in its parts, a perfect classification on such a basis is scarcely to be hoped for.

Chondritic aerolites. About eleven out of every twelve of the stony meteorites belong to a division to which Rose * has given the name of Chondritic (*chondros*, a grain): they present a very fine-grained but crystalline matrix or paste, consisting of olivine and enstatite or bronzite, with more or less nickel-iron, troilite, chromite, augite and anorthic felspar ; through this paste are disseminated round chondrules of various sizes and with the same mineral composition as the matrix; in some cases the chondrules consist wholly or in great part of glass.† In mineral composition they approximate more or less to terrestrial lherzolites. Of this division Knyahinya, Pane 4n. Pegu, Muddoor, Seres, Judesegeri, Khiragurh, Utrecht, and Nellore (pane 4p) afford good illustrations.

A carbonaceous group. Some meteorites belonging to this division are remarkable as containing carbon in combination with hydrogen and oxygen. Of these the Alais and Cold Bokkeveldt meteorites Pane 4n. are good examples: the former is combustible and has a bituminous smell; it contains also sulphates of magnesium, calcium, sodium and potassium, which can be dissolved out with water.

Aerolites without chondrules. 40. The remaining aerolites are not chondritic, and they Pane 4o. contain little or no nickel-iron; of these we may specially mention for their mineral composition the following :—

* Beschreibung und Eintheilung der Meteoriten. Berlin, 1864.
† Die mikroskopische Beschaffenheit der Meteoriten : von G. Tscher-mak. Stuttgart, 1883–5.

Juvinas, and *Stannern*, consisting essentially of anorthite and augite.

Petersburg, consisting of anorthite, augite and olivine, with a little chromite and nickel-iron : both Juvinas and Petersburg may be compared to terrestrial basalt.

Sherghotty, consisting chiefly of augite and maskelynite.

Angra dos Reis, consisting almost wholly of augite; olivine is present in small proportion.

Bustee, of diopside, enstatite and anorthic felspar, with some nickel-iron, oldhamite and osbornite.

Bishopville, of enstatite and anorthic felspar, with occasional augite, nickel-iron, troilite and chromite.

Roda, of olivine and bronzite.

Chassigny, consisting of olivine with enclosed chromite, and analogous in composition to a terrestrial dunite.

Is there a periodic recurrence?
41. The importance of the examination and classification of meteorites, with a view to a possible recognition of *periodicity* of fall of specimens presenting the same characters, need only be mentioned to be appreciated : such a determination is, however, rendered very difficult by the close similarity of structure and composition presented by the great majority of the aerolites of the large chondritic division.

Few aerolites are known which have not been seen to fall.
42. Attention has been already directed to the fact that although many meteoric irons, some of them like that of Cranbourne weighing several tons, have been found at various parts of the earth's surface, very few of them have been actually observed to fall : in the case of the stony meteorites just the opposite holds good, for they are never very large, and few are known which have not an authenticated date of fall. This may be due to the fact that a meteoric stone is less easily distinguished than is a meteoric iron from ordinary terrestrial bodies, and will thus in most cases remain unnoticed unless its fall is actually observed; while, further, a quick decomposition and disintegration must set in on exposure to atmospheric influences. The smaller size of the meteoric stones may be due to the greater ease with which they break up on the sudden increase of temperature of their outer surface, consequent

Separate stand.

on their entry into the earth's atmosphere. The largest meteoric stone known is one of those which fell at Knyahinya, Hungary, in 1866 : it weighs 647 lbs. and is preserved in the Vienna Museum.

The chondrules and their matrix.

43. If we now examine minutely the structure of the meteoric stones, it will be seen that almost all of them appear to be made up chiefly of irregular angular fragments, and that some of them bear a close resemblance to volcanic tuffs. In the large group of chondritic aerolites, chondrules or spherules, some of which can only be seen under the microscope while others reach the size of a cherry, are embedded in a matrix, apparently made up of minute splinters such as might result from the fracture of the chondrules themselves. In fact, until recently, it was thought by some * that the chondrules owe their form, not to crystallisation, but to friction, and that the matrix was actually produced by the wearing down of the chondrules through collision with each other either as oscillating components of a comet or during repeated ejection from a volcanic vent of some small celestial body. Chondrules have been observed, however, presenting forms and crystalline surfaces incompatible with such a mode of formation, and others have been described which exhibit features resulting from mutual interference during their growth.

The crystallisation of the chondrules is independent of their form, and must have started, not at the centre, but at various places on their surfaces; Sorby † argues that some at least of the chondrules must once have fallen as drops of fiery rain, and have assumed their shape in an atmosphere heated to nearly their own temperature. The chondritic structure is different from anything which has been observed in terrestrial rocks, and the chondrules are distinct in character from those observed in perlite and obsidian. After a minute study of the classical collection at Vienna, Brezina‡ lends his weighty support to the theory that the structural features of meteorites are the result of a hurried

* Pogg. Ann. 1858, vol. 105, p. 438 : Phil. Mag. 1876, ser. 5, vol. 1, p. 497.
† On the structure and origin of meteorites. *Nature*, 1877, vol. 15, p. 495.
‡ Die Meteoritensammlung d.k.k.min.Hofkabinetes in Wien. 1885, p. 19

crystallisation : and Wadsworth * accepts the same interpretation.

Some meteoric rocks appear to have been altered since their formation. **44.** Since the time of their consolidation some meteoric stones, as Tadjera, appear to have been heated throughout their mass to a high temperature : and in the case of Orvinio, Chantonnay, Juvinas and Weston, fragments are cemented together with a material having the same composition as the fragments themselves, thus giving rise to a structure resembling that of a volcanic breccia. Others seem to have experienced a chemical change, for some of the chondrules in Knyahinya and in Mezö-Madaras, when examined with the microscope, are found to be surrounded by spherical and concentric aggregations of minute particles of nickel-iron, perhaps due to the reducing action of hydrogen at a high temperature. Others, as Château-Renard, Pultusk, and Alessandria, present what in terrestrial rocks would probably be called faults : in some cases the fissures are seen to have been filled with a fused material after the chondrules have been broken and one side of the fissure has glided along the other. These peculiarities of structure suggest that the small body which reaches the earth is only a minute fragment of a much larger mass.

Do meteorites reach our atmosphere as clouds of gas or dust? **45.** The idea that meteorites arrive at our own atmosphere, not as fragments of rock, but as mere clouds of gas or dust, has been recently revived by Brezina. According to this hypothesis, the air, instead of dispersing the entering cloud, acts in the contrary way, and in a few seconds of time presses the particles together to form solid bodies. This idea is open to various objections, and in any case one can scarcely understand how large masses of iron, presenting a wonderful regularity of crystalline structure, can have been the result of so hurried a process : and if we once grant that the irons enter the atmosphere as solid bodies, it is difficult to believe that the same is not the case with the stones.

Where do meteorites come from? **46.** From the above it will be evident that the old theories that meteorites are ordinary stones struck by lightning, or carried to the sky by a whirlwind, or are concretions in the

* Lithological Studies. Cambridge, U.S.A. 1884, p. 110.

atmosphere, or are due to the condensation of a cloud coming from some volcano, or have been shot recently from terrestrial volcanoes, are inconsistent with later observation, and that the bodies reach our atmosphere from outer space. From what part or parts of space do they come? Their general similarity of structure and chemical composition, and more especially the presence of nickeliferous iron in almost every one, suggest that most, if not all of them, have had a common source, and that they are chips of a single celestial body.

Probably not from the sun, moon, earth, or other planet.

47. Sorby holds that they are probably ejected from the sun itself, though this is not easily to be reconciled with the fact that some of them are easily combustible. Others, among whom we may mention Laplace, have suggested that they come from volcanoes of the moon which are now active, but the suggestion, although mathematically sound, has no physical basis, for so far as one can discover, active volcanoes do not there exist: and Sir Robert Ball* has virtually excluded the lunar volcanoes which were active in times now long past, by pointing out that if a projectile from the moon once misses the earth, its chance of ever reaching the earth is too small to be worthy of mention. It has further been shown that, although the explosive force necessary to carry a projectile so far from one of the smaller planets that it will not return, is not very large, yet the initial velocity requisite to carry the body as far as the earth's orbit is so considerable, and the chance of hitting the earth so slight, that a more probable hypothesis is, to say the least, desirable. If these bodies have been shot from volcanoes of any planet, Sir Robert Ball is himself inclined, upon mechanical grounds alone, to believe that the projection was from our own in bygone ages; for as such projectiles, having once got away from the earth, would take up paths round the sun which would intersect the earth's orbit, every one of them would have a chance of some time or other meeting with the earth again at the point of intersection, and of appearing as a meteorite. The size and initial velocity requisite for the escape of a projectile through a lofty atmosphere would be enormous:

* Speculations on the source of Meteorites. *Nature*, 1879, vol. 19. p. 493.

even then the difficulty would still remain that meteorites generally, in their structure and material, differ from anything known to have been ejected from existing terrestrial volcanoes.

Nor is it probable that they are portions of a lost satellite of the earth, or are due to a collision of two planets, for in each of these cases we should expect to have received some of the larger fragments which must at the same time have been produced.

Much light is thrown on the history of meteorites by the discovery of a relationship with shooting stars and comets.

Shooting or falling stars. **48.** The meteorite-yielding fireball, referred to in Art. 17, is not the only luminous meteor, apart from lightning, with which we are acquainted. On a clear dark night any one can see a star shoot now and then across the firmament: it is estimated that on the average as many as fourteen are visible to a single observer every hour. Are the *shooting*, or, as they are often called, *falling stars* products of our own atmosphere, or do they, like the meteorites, come from outer space? In 1794 Chladni, in the memoir already referred to, gave reasons for believing that a meteoritic fireball and a shooting star are only varieties of one phenomenon.

The November star-showers. **49.** But long after the cosmic origin of meteorites had been generally acknowledged, the atmospheric origin of the shooting stars was still asserted, and it was not till the wondrous star-shower of 12–13th November, 1833,* that the cosmic origin of any of the shooting stars was finally established. During that night upwards of 200,000 shooting stars, according to a rough estimate, were seen from a single place; and the remarkable observation was made at various localities, widely distributed over North America, that the apparent paths of the shooting stars in the sky, when prolonged backwards, all passed through a point in the constellation Leo: this point of radiation appeared to rotate with the heavens during the eight hours for which the shower was visible.

Hence it was manifest that the star-shower was independent of the earth's rotation and must therefore have come

* *Olmsted.* American Jour. Sc., 1834, ser. 1, vol. 25, p. 363.

from outer space; that the radiation of the paths was only apparent and due to perspective; and that, relatively to an observer, the flights of all the shooting stars were really parallel to the direction of the apparent radiant point.

On the same date in the three following years the shower was repeated though on a less grand scale, and the constancy of the radiant point was confirmed : similar small showers had been seen also in 1831 and 1832 before the radiation had been noticed. Though in the years immediately before and after 1831–6 no remarkable display of November meteors took place, it was remembered that a similar shower had been chronicled by Humboldt and by Ellicott, as observed by them on 12th November, 1799; and a study of ancient records revealed the fact that a grand star-shower had attracted general attention at intervals of 33 years ever since A.D. 902, though the date had steadily advanced during that long period from the middle of October to the middle of November. The only sufficient explanation of the observed facts is that a swarm of isolated small bodies, solid and non-luminous—meteorites in fact—is moving in an orbit round the sun; the orbit intersects that of the earth, and the earth meets the swarm at the place of intersection. The swarm can be only a few hundred thousand miles thick, for the earth, travelling through space at the rate of 66,000 miles an hour, passes through the densest part in 2 or 3 hours, and through the whole in 10 to 15 hours; its length, however, must be enormous, amounting to hundreds of millions of miles, for although the meteorites move with a velocity comparable with that of the earth, the swarm takes 5 or 6 years to pass the place of intersection with the earth's orbit, thus causing star-showers, more or less dense, during that number of years. The isolated bodies or meteorites become luminous, as already explained in Art. 17, owing to their entry into the earth's atmosphere.

Schiaparelli has shown that the unequal attraction of the sun for the individuals of a swarm of meteorites moving round it would scatter them along the orbit, and in the course of time produce a more or less complete ring; if this intersects the earth's orbit an annual star-shower must ensue.

The August star-shower and its comet.

Star-showers related to comets.

Comets.

50. A small annual star-shower occurs, in fact, on 10–11th August,* and has been observed since A.D. 830: it radiates from a point in the constellation Perseus. Schiaparelli calculated in 1866 the orbit and motion of the meteorites producing it, and was surprised to find that the numbers corresponded exactly with those calculated for one of the recently observed comets; in other words, a comet was moving in the path of the meteorites, and at exactly the same speed. At the same time Schiaparelli gave numbers defining the motions of the meteorites which would cause the periodic November star-showers.

51. Immediately afterwards, when the numbers calculated by Oppolzer for the orbit of the comet discovered by Tempel were published, it was seen that they were really identical with those already calculated by Schiaparelli for the orbit of the meteorites of the November star-shower, and that here again a comet and a swarm of meteorites were moving in exactly the same path at exactly the same rate.

Almost immediately afterwards it was shown that the radiant points of the small star-showers of April 20–21st and November 27–28th both correspond to the orbits of known comets.

It was evident that these could not be accidental coincidences, and that comets and their attendant swarms of meteorites are closely related to each other.

52. An intimate connection between, if not complete identity of, meteorites, shooting stars, and comets, had indeed long been suspected. Astronomers were convinced that comets, though occasionally of enormous size, are always of extremely small mass, since they pass by the earth and other planets without sensible disturbance of their motions; the comet of 1770 passed through the system of Jupiter's satellites without any perceptible action upon them: it has been calculated that the mass of a small comet may be about eight pounds. Again, the light of a comet, like that of a cloud or planet, was seen to be partially polarised: hence part, at least, must be reflected sunlight, for the plane of polarisation passes through the sun's place. Further, stars

---

* Report Brit. Assoc., 1868, p. 394.

of very small magnitude have been seen not only through the tail, but even through the nucleus, of a comet without any apparent alteration of position by refraction : hence it was inferred that a comet is not a continuous mass, but consists of particles so far distant from each other that a ray of light may pass through the comet without meeting a single one of them. Such a constitution likewise accounts for the absence of phases of the reflected light : for although only half of each particle will be directly illuminated by the sun, the remaining half will receive light irregularly reflected from the particles more distant from the sun.

Among others, Chladni in 1817 had referred to the great similarity in the motions of comets and meteorites : Olmsted, in 1834, had calculated the orbit of a comet which would cause the November star-shower ; his results were wrong owing to the assumption that the shower was annual : Cappocci, in 1842, gave reasons for believing that a meteorite is a small comet : Reichenbach, in 1858, in a most elaborate paper,* sought to prove that a comet is a swarm of meteorites ; that each chondrule of a meteorite had once been an individual of a cometary swarm, and owes its rounded shape to frequent collision with its fellows ; that the rest of the stone consists of the broken splinters thus produced ; and that the brecciated aspect of many meteorites is due to collisions in the denser part or nucleus of a comet. As already pointed out in Art. 43, later modes of investigation have led petrologists to reject this explanation of the rotundity of the chondrules.

Other star-showers. **53.** In addition to the few radiant points which correspond to swarms moving in orbits identical with those of known comets, there are numerous radiant points which have not yet been recognised as related to existing comets, and may possibly be due to swarms produced by the dispersal of comets along their orbits ; but there are others of which there is yet no satisfactory explanation. A cometary swarm is thin, and is passed through in a few hours ; the stars are only seen to radiate from the corresponding point of the sky for the same short time : but there are other radiant points

* Pogg. Ann., 1858, vol. 105, p. 438.

which have a duration of several months, and this is the case notwithstanding the constantly changing direction of the earth's motion in space.* Since the position of the radiant point in the sky as seen by a terrestrial observer depends not only on the direction in which the swarm is moving, but also on the velocity and direction of motion of the observer through space, it is easily seen that a radiant point having a fixed position during some months corresponds to something quite distinct from a cometary swarm.

The breaking up of comets.
54. The history of Biela's comet † is of great interest as throwing light on the relationship of comets and swarms of meteorites. Though already observed in 1772 and in 1806, this comet was not recognised as periodic till it was seen by Biela in 1826, when its orbit was determined. On its returns in 1832 and 1845 it was found in its calculated positions, but in the latter year was seen to be double, a small comet being visible beside a larger one. Vast changes took place during the time the companions were visible. The smaller one grew both in size and brightness, each threw out a tail, the smaller threw out a second tail, afterwards the larger showed two nuclei and two tails, then the smaller became the brighter of the two companions; next three tails were shown by the primary, and three cometary fragments were visible round its nucleus. On the next return, in 1852, the two comets were farther apart, one being more than a million miles ahead of the other. The next favourable return was to be in 1866, and the orbit was now so well known that the positions of the two companions could be calculated beforehand with great precision; owing to the changes which had been visibly taking place, the arrival of the comets was looked forward to with great interest by astronomers. Neither in 1866, nor on the next occasion in 1872, were they to be seen in their calculated positions, and a careful examination of the whole sky failed to lead to their discovery.

The connexion between several comets and meteoritic swarms having in the meantime been established, it was

---

* Denning. *Nature*, 1885, vol. 31, p. 463.
† Newton. *Nature*, 1886, vol. 33, pp. 392, 418.

now surmised that Biela's comet might have been scattered along part of its path, and that some evidence of the dispersal might perhaps be obtained on the next occasion, November 27th, 1872, of the passage of the earth across the comet's orbit. In fact the star-shower of that date, with a radiant point corresponding to the orbit of Biela's comet, was observed to be much more dense than usual, the stars shooting across the sky at the rate of a thousand an hour for several hours.

Passage of the earth through a comet. 55. Klinkerfues, a German astronomer, was struck with the idea that if this star-shower were really due to the passage of the earth through a moving swarm of meteorites, the latter might possibly be visible as it departed from our neighbourhood. The swarm having come from a radiant point in the northern sky, after passing the earth would need to be sought near the opposite point in the southern sky; he telegraphed, therefore, to the Madras observatory, asking Pogson, the astronomer, to search for the swarm in the direction opposite to the radiant point. The search was successful; on two mornings a small comet was distinctly seen, and on the second morning it showed a tail with an apparent length equal to one-fourth the apparent diameter of the moon. Bad weather came on, and the comet got away without being again seen. The two Madras observations agree with a motion in the orbit of Biela's comet, and show that the earth had passed excentrically through the small comet seen by Pogson. This small comet was probably a third fragment of Biela's, for it was 200 million miles behind the calculated position of the first two. From this observation it is inferred that a swarm of meteorites, though only manifesting itself by a star-shower when passing through the earth's atmosphere, at some distance from us may be visible as a comet by reflected sunlight.

Fall of a meteorite during a star-shower. 56. A dense star-shower * recurred on the same date, 27th of November, in 1885, the principal part being over in six hours. The hourly number visible at one place at the time of greatest density was estimated at 75,000. In the densest part of the stream, the average

* *Newton.* American Jour. Sc., 1886, ser. 3, vol. 31, p. 409.

distance of the individuals from each other was about twenty miles.

During this star-shower a piece of iron weighing about 8 lbs. was seen to fall at Mazapil in Mexico : * in external characters and chemical composition it is similar to the other meteoric irons : the simultaneity was probably accidental.

57. It may be asked why, if star-showers are caused by the entry of solid bodies into our atmosphere from without, there is only one authentic instance of a solid being actually seen to fall and being picked up during such a shower. It being absolutely beyond question that star-showers do come from outer space, we can only seek an explanation in the size or speed of the entering individuals, or in the nature of their material. A sufficient reason is to be found in the small size of the individuals; for the meteorites which actually reach the ground rarely weigh more than a few pounds, and are often quite minute ; a small diminution of the original individual would thus ensure its complete destruction before the planetary velocity was exhausted : that the individuals of a swarm are extremely minute follows from the fact that the total mass of the biggest swarm is small, while the number of the individuals seems almost infinite.

58. Between the small silent shooting star visible only in the telescope and the large detonating meteorite-yielding fireball there is every gradation; during the star-showers themselves many fireballs of great size and brilliancy are seen, while the smaller individuals appear in no way different from the solitary shooting star. The luminous meteors, large and small, are in the upper atmosphere, few higher than 100 miles, few lower than 30 miles from the earth's surface ; they all have velocities of the same order of magnitude, comparable with that of the earth in its orbit ; in each there must be a solid body, as is proved by the long path in the sky, attendant gas or vapour would be immediately blown away or burnt ; large and small present similar varieties of colour, and leave similar luminous trains ; examination with the spectroscope teaches us that the light of the meteors is such as would result from the ignition of such meteorites

* *Hidden.* American Jour. Sc., 1887, ser. 3 vol. 33, p. 223.

as have actually reached the ground. The frequent absence of detonation may likewise be due in many cases to the small size of the entering meteorite.

The light of a comet. 59. That part of the light of a comet is reflected sunlight is confirmed by examination with the spectroscope, in which instrument is seen a feeble continuous spectrum crossed by dark lines, identical with those afforded by the direct light of the sun. But a comet is also more or less self-luminous, for in addition to the continuous spectrum, there are bright flutings and bright lines to which much attention has been given. The three ordinary bright flutings were found by Huggins in 1868 to be identical with the spectrum obtained when an electric spark is passed through olefiant gas, and they are now recognised as due to carbon.

Tait's suggestion. 60. The discovery made by Schiaparelli proves, as already pointed out, that there is a relationship between comets and meteoritic swarms; Schiaparelli himself held the view that a comet and its attendant swarms are merely of identical origin. In 1869 * Tait discussed, from a purely dynamical point of view, the question as to whether the swarm of meteorites attending a comet may not really be part of the comet itself; he shewed that many cometary characters can be mechanically explained on the assumption that comets are really swarms of small meteorites, and pointed out that the self-luminosity may be produced by the heating of the individuals through collision with each other.

Reproduction of the spectrum of a comet. 61. Flutings exactly identical with those seen in the spectrum of a comet were obtained by Wright in 1875 † on allowing the electric glow to pass through a heated tube, in which, after the introduction of fragments of the Iowa meteorite, the gaseous density had been reduced by an air-pump. The bright lines, too, in the spectrum of a comet, even when nearest to the sun, are found by Lockyer to be identical with those yielded when the electric glow is passed over ordinary meteorites at comparatively low temperatures; and further, the changes in these lines as the comet approaches and recedes from the sun are exactly those which

* Proc. Roy. Soc., Edinb., 1869, vol. 6, p. 553.
† American Jour. Sc., 1875, ser. 3, vol. 10, p. 44.

take place on variation of the temperature of the meteorites enclosed in the glow-tubes.

A comet is a swarm of meteorites. 62. From these facts it is inferred that a comet is in every instance a swarm of isolated meteorites, at a not very high temperature, shining partly by reflected sunlight and partly by the electric glowing of the gases evolved owing to the action of the sun's heat on the meteorites: further, some of the heat may be due to the clashing together of the meteorites, the grouping of which becomes more and more condensed as the swarm approaches the sun.

The gases driven from the meteorites would be quite sufficient in quantity to form the tail of the comet: as pointed out by Wright, a meteorite like that which fell at Cold Bokkeveldt would furnish 30 cubic miles of gas measured at the pressure of our own atmosphere, and in space itself this gas would expand to enormous dimensions owing to the small mass and attraction of the meteoritic swarm. We are still uncertain, however, as regards the actual physical condition of the matter composing the tail of a comet.

Saturn's rings are probably swarms of meteorites. 63. Clerk-Maxwell proved so long ago as 1857 that the stability of the rings which revolve round the planet Saturn is inconsistent with their being formed of continuous solid or liquid matter; and has shown, by mechanical reasoning, that they must be revolving clouds of small separate bodies, like cannon-shot, each moving as a satellite and almost independent of the rest in its motion: observation with the spectroscope supports this conclusion.

Nebulæ. 64. Tait,* in 1871, going still farther into space, suggested that the nebulæ may likewise be clouds of meteorites, and pointed out that the heat produced by the clashing of the ndividuals of such an immense group as a nebula evidently is would be quite adequate for the production of their light. Reichenbach, in 1858, before the self-luminosity had been proved by means of the spectroscope, imagined a nebula to be a cloud of isolated meteorites, illuminated by some neighbouring sun : Chladni supposed a nebula to be a cloud of phosphorescent dust. Lockyer now shows that the

* Proc. Roy. Soc., Edinb., 1871, vol. 7, p. 460.

D

bright lines (generally accompanied by a certain amount of continuous spectrum) which have been observed in nebular spectra, and had led to the nebulæ being regarded as masses of glowing gas, are consistent with this view, for they are closely related to the low temperature lines obtained when a gentle electric glow is passed over meteorite-fragments in a tube containing gases given out by them, and of which the density has been reduced by the air-pump; further he points out that the nebular spectrum is identical with that of the comets of 1866 and 1867 when distant from the sun. Hence in all probability a nebula and a comet are of identical constitution, and a comet is merely a nebula which has become entangled in the solar system.

Stars. 65. The examination and classification of the spectra of the stars has likewise led to remarkable conclusions. Secchi, following Rutherfurd, found that the stars could be distributed into classes according to the characters of their spectra,* and his classification has since, with little modification, been adopted by Vogel and Dunér, by whom several thousand star-spectra have now been systematically mapped. The first three classes are characterised by absorption, the fourth by radiation.

In the spectra of Class I. the absorption is small and simple, the dark lines being broad and few; the stars themselves are white: here belong Sirius and Vega.

In Class II. the dark lines are thinner and more numerous; the stars are bluish-white to reddish-yellow: to this class belong the Sun, Arcturus, Capella.

The absorption in Class III. manifests itself predominantly as flutings, though there are also many thin lines: the stars are orange or red: in one division (*a*) of this class the darkest part and the sharpest edge of each fluting is towards the violet end of the spectrum, as in Betelgeux; in a smaller division (*b*) the darkest part of each fluting is towards the red end, as in star 152 Schjellerup; the fluting absorption of the latter division being due to carbon.

The remaining Class IV. is an extremely small one: the spectra are characterised by bright lines: those of one divi-

---

* Lockyer. *Nature*, 1886, vols. 33 and 34.

sion (*a*) show hydrogen lines, and the stars are of a blood-red colour: in the other division (*b*), consisting as yet of only about six stars, the hydrogen lines are absent.

Supposed cooling of all the stars.

66. Soon after the classification suggested by Secchi had been announced, it was surmised that the differences in the stars of the first three classes might be due, not so much to differences of matter, as to differences of temperature, and that a very hot star such as, from its brightness and distance, its small and simple absorption, and the development of the blue end of its spectrum, Vega is believed to be, would, on getting older and colder, pass from Class I. to Class II., and thence to one or other of the divisions of Class III.

New stars.

67. In 1866 a star of 9th or 10th magnitude burst into greater brilliancy and nearly reached the intensity of Vega; analysis of its spectrum showed that the increase of brilliancy was due to hydrogen. Almost as suddenly the light went down again, and within a month returned to its original brightness. Ten years later, another new star of the 3rd or 4th magnitude appeared at a place in the sky where no star had been noticed before; its spectrum showed numerous bright lines; gradually in the course of a year it dwindled down to the 10th magnitude, then giving the telescopic appearance and the spectrum of a nebula.

The appearance of a new star has been generally attributed to the collision of two bodies in space; Lockyer * has pointed out that the rapidity of the change in the brilliancy, so different from that of other stars, may be due to the smallness of the mass, and that such a star may be produced by the collision of two swarms of widely separated meteorites. He has lately shown that the changes in the spectrum as such a star varies in brightness are confirmatory of this view.

The heat of the sun.

68. That the heat of our own sun was produced by the impact of matter in past times is now generally acknowledged;† for the only other conceivable natural explanation,

* *Nature*, 1877, vol. 16, p. 413.
† Treatise on Natural Philosophy, by Thomson and Tait: *Cambridge*, 1883, vol. 1, part 2, p. 487.

combustion, is quite insufficient; the greatest amount of heat obtainable from the most advantageous chemical combination of any known elements, having a total mass equal to that of the sun, would not cover the sun's expenditure for more than three thousand years, while there is no difficulty on the meteoritic explanation in providing a supply of heat sufficient to cover the loss by radiation during 20,000,000 years.

The present loss of the sun's heat by radiation is probably not covered by the fall of bodies into the sun, since the requisite mass would, if from distant regions, visibly affect the motions of the planets by its attraction, and even if circulating round the sun at no great distance from it would seriously disturb the motions of some of the comets.

Evolution of the heavenly bodies.

69. By a careful study of the spectra at various temperatures of the elements and compounds found in those meteorites which have reached our earth and been preserved, Lockyer [*] has been led to infer that the stars are not at present all cooling down, but on the contrary are mostly rising in temperature, and, like the nebulæ, are constituted of separate meteorites in continual relative motion, and rendered hotter and hotter through contraction of the grouping and transformation of the energy of position and motion into heat. This increase of temperature must continue during successive ages, until the energy of position and motion of the separate meteorites is wholly transformed, the separate masses having then combined to form a single white hot body which will gradually cool down to the state in which our own moon now is. If a swarm of meteorites forming one nebula be subjected to the external action of another moving swarm of meteorites, intermediate stages resembling the conditions of Saturn and of the solar system will ensue.

According to this spectroscopic affirmation of the nebular theory, all the heavenly bodies are constituted of the same kinds of elementary matter, those in fact which are found in meteorites and our own earth, and the difference is solely due to temperature; a nebula in its gradual passage to the

* Proc. Royal Society, 1887, vol. 43, p. 117 : 1888, vol. 44, Bakerian lecture.

lunar condition showing every phase of spectrum observed in the stars as now existent.

Meteorites present no evidence of life.   **70.** Finally, it may be asked whether or not meteorites bring us any tangible evidence of the existence of living beings outside our own world. To this we may briefly answer, that while an organic origin can scarcely be claimed for the graphite present in the meteoric irons, there are no less than six meteoric stones which contain, though in very minute quantity, carbon compounds of such a character that their presence in a terrestrial body would be regarded as doubtlessly an indirect result of animal or vegetable existence. On the other hand, the stony matter is such that in a terrestrial body an igneous origin would be assumed.

Professor Maskelyne points out that these carbon compounds can be completely removed without a preliminary pulverisation of the stone, and thus seem to be contained merely in the pores; he suggests that they may have been absorbed by the stones in their passage through an atmosphere containing the compounds in a state of vapour. In any case, it is impossible to prove that there is a necessary relation between these compounds of carbon and the existence of living beings.

Chondrules have been mistaken for organisms.   **71.** In 1880 * descriptions were given of sponges, corals, crinoids and plants, found in several meteorites, chiefly in that of Knyahinya, but the memoir has been generally regarded as an elaborate jest. The chondrules with their excentrically radiating crystallisation are there classified and named as sponges, corals and crinoids, while the structure of meteoric iron, revealed by the Widmanstätten figures, is regarded as a result of plant life. There can be no hesitation in asserting that as yet no organised matter has been found in meteorites.

   * Die Meteorite (Chondrite) und ihre Organismen: von Dr. O. Hahn. Tübingen, 1880.

LIST OF THE METEORITES

REPRESENTED IN THE COLLECTION.

The references in the second column correspond with numbers and letters on the cases, and indicate the pane behind which the meteorite will be found.

Weights under one gram are not giveñ. 1,000 grams are equivalent to 2·205 avdp. lbs.

I. SIDERITES

or Meteoric Irons,

consisting chiefly of nickeliferous iron, and enclosing schreibersite, troilite, graphite, &c.).

A. FALL RECORDED.
[Arranged chronologically.]

No.	Pane.	Name of Meteorite and Place of Fall.	Date of Fall.	Weight in grams.
1	1c	**Agram** (Hraschina), Croatia, Austria.	May 26, 1751.	282·3
2	1c	**Charlotte**, Dickson County, Tennessee, U.S.A.	July 31, or Aug. 1, } 1835.	77·5
3	1c,4l	**Braunau** (Hauptmannsdorf), Bohemia.	July 14, 1847.	553·2
4	1c,4l	**Victoria West**, Cape Colony, South Africa.	Fell in 1862.	158·5
5	1c,4h	**Nedagolla**, Murangi, Vizagapatam, Madras, India.	Jan. 23, 1870.	4,379·7
6	1c	**Rowton**, near Wellington, Shropshire.	April 20, 1876.	3,109·0
7	1c	**Mazapil**, Zacatecas, Mexico.	Nov. 27, 1885.	14·0
8	1c	**Cabin Creek**, Johnson County, Arkansas, U.S.A.	March 27, 1886.	5·2

B. FALL NOT RECORDED.

[Arranged geographically.]

No.	Pane.	Name of Meteorite and Place of Find.	Report of Find.	Weight in grams.
9	1c	**Newstead,** Roxburghshire, Scotland. Found in 1827, three or four feet deep in a stratum of clay: its meteoric origin was recognised by Dr. J. A. Smith in 1862.	Edinb. New Phil. Journ. New Ser., 1862, vol. 16, p. 108.	8,129·0
10	1c	**La Caille,** near Grasse, Alpes Maritimes, France. For about two centuries it was in front of the church of La Caille and was used as a seat: its meteoric origin was recognised by Brard in 1828.	Acad. Sci. Bordeaux, 1829, p. 39.	375·0
11	1c	**S. Julião de Moreira,** Ponte de Lima, Minho, Portugal. Known since 1883: described by Ben-Saude in 1888.	Comm. da commiss. d. trab. geol. de Portugal, 1888, vol. 2, p. 14.	9·1
12	1p	**Obernkirchen,** near Bückeburg, Schaumburg-Lippe, Germany. Found in a quarry on the Bückeberg 15 feet below the surface, and thrown aside: recognised as meteoric by Wicke and Wöhler, in 1863.	Pogg. Ann. 1863, vol. 120, p. 509.	35,366·5
13	1c	**Bitburg,** Rhenish Prussia. Dug up about 1807, taken to Trèves and put into a furnace: afterwards thrown away with the waste: later, fragments of it having been recognised by Gibbs as meteoric, the mass was searched for by Nöggerath and re-discovered in 1824.	Schweigg. Journ. 1825, vol. 43, p. 1.	1,349·0
14	1c	**Nauheim,** Giessen, Ober-Hesse, Germany. Found in 1826; reported by Wille in 1828.	Geognost. Beschr. Taunus- u. Vogelsgebirge; von G. A. Wille. Mainz, 1828, p. 51.	3·6
15	1d, 4l	**See-Läsgen,** Brandenburg, Prussia. Found in draining a field: several years afterwards, in 1847, it was met with by Hartig and recognised as meteoric.	Pogg. Ann. 1848, vol. 73, p. 329; 1849, vol. 74, p. 57.	9,846·5
16	1d	**Schwetz,** Prussia. Found in 1850 in making a road; it was about 4 feet below the surface: described by Rose in 1851.	Pogg. Ann. 1851, vol. 83, p. 594.	1,062·5
17	1d	**Nenntmannsdorf,** Pirna, Saxony. Found in 1872 about 2 feet below the surface: reported by Geinitz in 1873.	Sitzungs-Ber. d. n. G. Isis in Dresden, 1873, p. 4.	15·6

No.	Pane.	Name of Meteorite and Place of Find.	Report of Find.	Weight in grams.
18	1d	**Tabarz**, near Gotha, Germany. Said to have been seen by a shepherd to fall on Oct. 18, 1854 : described in 1855 by Eberhard, to whom the rust seemed incompatible with a recent fall.	Ann.Chem. Pharm. 1855, vol. 96, p. 286.	9·0
19	1d	**Elbogen**, Bohemia. Preserved for centuries at the Rathhaus of Elbogen : its meteoric origin was recognised by Neumann in 1811.	Gilb. Ann. 1812, vol. 42, p. 197.	94·8
20	1d	**Bohumilitz**, Prachin, Bohemia. Laid bare by heavy rain in 1829.	Verh. Ges. Mus. Böhm. April 3, 1830, p. 15.	118·5
21	1d	**Lénárto**, Sáros, Hungary. Found in 1814 : described by Tehel in 1815.	Gilb. Ann. 1815, vol. 49, p. 181.	2,028·5
22	1e	**Arva** (Szlanicza), Hungary. Made known by Haidinger in 1844.	Pogg. Ann. 1844, vol. 61, p. 675.	9,010·7
23	1d	**Nagy-Vázsony**, Veszprim, Hungary. Found in 1890 : described by Brezina in 1896.	Ann. d.k.k.Naturh. Hofmus. Wien, 1896, vol. 10, pp. 284, 356.	69·7
24	1d	**Tula** (Netschaëvo), Russia. Found in 1846 in making a road : it was 2 feet below the surface : recognised as meteoric by Dr. Auerbach in 1857.	Wien. Akad. Ber., 1860, vol. 42, p. 507.	1,076·8
25	1e	**Sarepta**, Saratov, Russia. Found in 1854 : reported by Auerbach in the same year.	Bull. Soc. Nat. Moscow, 1854, p. 504.	296·0
26	1d	**Verkhne-Dnieprovsk**, Ekaterinoslav, Russia. Found in 1876.		24·8
27	1e	**Bischtübe**, Nikolaev, Turgai, Russia. Found in 1888: described by Kislakovsky in 1890.	Bull de la Soc. Imp. des Natur. de Moscou, 1890, p. 187.	88·0
28	1d	**Petropavlovsk** (gold washings), Mrasa River, Tomsk, Asiatic Russia. Found about 32 feet from the surface : given to the Director of the Kolyvani Works in 1841 and described by Sokolovskji in the same year.	Erman's Archiv f. wiss. Kunde von Russland, 1841, vol. 1, p. 314.	12·0

No.	Pane.	Name of Meteorite and Place of Find.	Report of Find.	Weight in grams.
29	1e	**Taiga,** Krasnojarsk, Jenisseisk, Asiatic Russia. Found in 1890 (Siemachko).		10·0
30	1d	**Ssyromolotovo,** Keshma, Jenisseisk, Asiatic Russia. Known since the year 1873 : described by Göbel in 1874.	Bull. Ac. Imp. des Sc. de St. Petersb. 1874, vol. 19, p. 544.	3·8
31		**Verkhne-Udinsk** (Niro river), Transbaikal, Asiatic Russia. Found in 1854 : noted by Buchner in 1865.	Pogg. Ann. 1865, vol. 124, p. 599.	2,904·0
32	1b	**Nejed** (Wanee Banee Khaled), Central Arabia. Said to have been seen to fall in 1863 ; probably this is a mistake and the time of fall unknown : described by L. F. in 1887.	Mineralog. Magaz. 1887, vol. 7, p.179.	59,420·0
33	1e	**Great Fish River** (east bank of), Western part of South Africa. Reported by Alexander in 1838. {Through a mistake of Partsch this locality has been confused with that of the Cape of Good Hope iron}. L. F.	An Exped. of Dis. Inter. Africa (countries of the Great Namaquas, Boschmans, and Hill Damaras) : by Sir J. E. Alexander : 1838, vol. 2, Appendix, p. 272.	20·4
34	1e	**Great Namaqualand** (north of the Orange River), South Africa. {Found at some distance in Namaqualand, and brought down to the Orange River ; long afterwards, about 1860, it was removed by Mr. Wild to Cape Town}. L. F.		1,440·0
35	1e	**Orange River,** South Africa. Described by Shepard in 1856.	Amer. Jour. Sc. 1856, ser. 2, vol. 21, p. 213.	98·0
36	1e	**Springbok River,** Namaqualand, South Africa. *From Dr. H. J. Burkart's Collection.*		9·5
37	1e	**Lion River,** Great Namaqualand, S. Africa. Found on a clay plain: described by Shepard in 1853. {Some of the above Namaqualand masses may have been transported from the same locality}. L. F.	Amer. Jour. Sc. 1853, ser. 2, vol. 15, p. 1.	390·0

No.	Pane.	Name of Meteorite and Place of Find.	Report of Find.	Weight in grams.
38	1e	**Hex River Mountains,** Cape Colony, South Africa. Found in 1882 : described by Brezina in 1896.	Ann. d.k.k. Naturh. Hofmus. Wien, 1896, vol. 10, pp. 291, 349.	245·0
39	1e	**The Cape of Good Hope iron:** found at a distance of about 15 English miles from the coast, between Karega and Kasuga rivers, Bathurst, Cape Colony, South Africa. Found in 1793 : mentioned in 1801 in 'Barrow's Travels,' vol. i. p. 226: full particulars were given in 1804 by Van Marum.	Natuur. Maatsch. Wetensch. Haarlem, 1804, vol. 2, p. 258.	328·7
40	1e	**Kokstad,** Griqualand East, South Africa. Known since 1887 : described by Brezina in 1896.	Ann d.k.k. Naturh. Hofmus. Wien, 1896, vol. 10, pp. 284, 351.	203 0
41	1e	**St. Augustine's Bay,** Madagascar. The existence of iron in Madagascar was made known in 1845.	Buchner's Meteoriten, p. 171.	5·6
42	1e	**Prambanan,** Surakarta, Java. Known as early as 1797, and probably earlier : described by Baumhauer in 1866.	Arch. Néer. Haarlem, 1866, vol. 1, p. 465.	8·9
43		**Thunda,** Windorah, Diamantina District, Queensland, Australia. Described by Liversidge in 1886.	Jour. and Proc. Roy. Soc. of New South Wales, 1887, vol. 20, p. 73.	396·0
44	1f	**Cowra,** Bathurst, New South Wales.		192 0
45	Sep. Stand,	**Cranbourne,** near Melbourne, Victoria, Australia.	Wien. Akad. Ber. 1861, vol. 43, Abth. 2, p. 583.	3,731,000·0
	1f	Known since 1854: described by Haidinger in 1861.		
	1f	{Fragments found in Abel's collection of minerals with the label " Yarra Yarra River—Date 1858 " had probably been detached from one of the two masses of Cranbourne}. L. F.		214·0
46	1f	**Youndegin,** 70 miles E. of York, Western Australia. Found in 1884: described by L. F. in 1887.	Mineralog. Mag. 1887, vol. 7, p. 121.	13,157·0
47	1f	**Madoc,** Hastings County, Ontario, Canada. Found in 1854 : described by Hunt in 1855.	Amer. Jour. Sc. 1855, ser. 2, vol. 19, p. 417.	216·0

No.	Pane.	Name of Meteorite and Place of Find.	Report of Find.	Weight in grams.
48	1*f*	**Welland,** Ontario, Canada. Ploughed up in 1888 : described by Howell in 1890.	Proc. Rochester Ac. of Sc. 1890, vol. 1, p. 86.	466·0
49	1*f*	**Iron Creek,** Battle River, North Saskatchewan, Canada. Removed about 1869 : described by Coleman in 1886.	Proc. and Trans. Roy.Soc.of Canada, 1887, vol. 4, sec. 3, p. 97.	79·5
50	1*f*	**Scriba,** Oswego County, New York, U.S.A. Dug up about 1834 and given to a blacksmith : described as meteoric by Shepard in 1841.	Amer. Jour. Sc. 1841, ser. 1, vol. 40, p. 366 ; 1847, ser. 2, vol. 4, p. 75.	132·3
51	1*h*	**Lockport** (Cambria), Niagara County, New York, U.S.A. Turned up by plough : described as meteoric by Silliman in 1845.	Amer. Jour. Sc. 1845, ser. 1, vol. 48, p. 388.	5,329·0
52	4*l*	**Seneca River,** Cayuga County, New York, U.S.A. Found in 1851, when digging a ditch : described by Root in 1852.	Amer. Jour. Sc. 1852, ser. 2, vol. 14, p. 439.	54·5
53	1*l*,4*l*	**Burlington,** Otsego County, New York, U.S.A. Turned up by plough some time previous to 1819, and described by Silliman in 1844.	Amer. Jour. Sc. 1844, ser. 1, vol. 46, p. 401.	290·0
54	1*g*	**Pittsburg** (Miller's Run), Alleghany County, Pennsylvania, U.S.A. Described by Silliman in 1850 : date of find unknown.	Proc. Amer. Assoc. for the year 1850, vol. 4, p. 37.	208·5
55	1*g*	**Emmittsburg,** Frederick County, Maryland, U.S.A. Found in 1854.		6·6
56	1*l*	**Staunton,** Augusta County, Virginia, U.S.A. Five specimens have been found. Three specimens, of which two at least were found in 1869, were described by Mallet, in 1871. A fourth was found about 1858-9, thrown away, used in the construction of a stone fence, then as an anvil; was next built into a wall : in 1877 it was taken out, and its meteoric nature was recognised by Mallet. A fifth was described by Kunz in 1887.	Amer. Jour. Sc. 1871, ser. 3, vol. 2, p. 10. Amer. Jour. Sc. 1878, ser. 3, vol. 15, p. 337. Amer. Jour. Sc. 1887, ser. 3, vol. 33, p. 58.	2,796·8

No.	Pane.	Name of Meteorite and Place of Find.	Report of Find.	Weight in grams.
57	1*h*	**Greenbrier County** (near the summit of the Alleghany Mountain, 3 miles north of White Sulphur Springs), West Virginia, U.S.A. Found about 1880 : described by L. F. in 1887.	Mineralog. Mag. 1887, vol. 7, p. 183.	2,236·0
58	1*h*	**Jenny's Creek**, Wayne County, West Virginia, U.S.A. The first piece was found before the Spring of 1883 and lost sight of; two other pieces were found in 1883 and 1885 respectively : reported by Kunz in 1885.	Proc. Amer. Assoc. for the year 1885, vol. 34, p. 246.	78·0
59	1*f*	**Smith's Mountain**, Rockingham County, N. Carolina, U.S.A. Reported by Genth in 1875 to have been found in 1866.		77·3
			Rep. Geol. Surv. N. Carolina, by Kerr : *Raleigh*, 1875, vol. 1, app. C, p. 56.	
		Reported by Smith in 1877 to have passed into the hands of Kerr about 1863.	Amer. Jour. Sc. 1877, ser. 3, vol. 13, p. 213.	
		No mention of date of find by Genth when describing the meteorite in 1885.	Min. and Min. Loc. of N. Carolina: *Raleigh*, 1885, p. 15.	
60	1*f*	**Guilford County**, N. Carolina, U.S.A. Date of find unknown : first described by Shepard as terrestrial in 1830, but in 1841 its meteoric origin was recognised by him.	Amer. Jour. Sc. 1830, ser. 1, vol. 17, p. 140; and 1841, vol. 40, p. 369.	15·0
61	1*g*	**Lick Creek**, Davidson County, North Carolina, U.S.A. Found in 1879 : described by Hidden in 1880.	Amer. Jour. Sc. 1880, ser. 3, vol. 20, p. 324.	20·0
62	1*k*	**Linnville Mountain**, Burke County, N. Carolina, U.S.A. Found about 1882 : described by Kunz in 1888.	Amer. Jour. Sc. 1888, ser. 3, vol. 36, p. 275.	21·2
63	1*h*	**Bridgewater**, Burke County, N. Carolina, U.S.A. Found by a ploughman : described by Kunz in 1890.	Amer. Jour. Sc. 1890, ser. 3, vol. 40, p. 320.	51·0
64	1*h*,4*l*	**Jewell Hill**, Walnut Mtns., Madison County, N. Carolina, U.S.A. One was given to Smith in 1854, and described by him in 1860.	Amer. Jour. Sc. 1860, ser. 2, vol. 30, p. 240; and Orig. Res. in Min. and Chem. by Lawrence Smith, 1884, p. 409.	130·2

No.	Pane.	Name of Meteorite and Place of Find.	Report of Find.	Weight in grams.
65	1*h*	A second was found in use in 1873, supporting a corner of a rail-fence : described as from Duel Hill by Burton in 1876. The etched figures are different for the two masses.	Amer. Jour. Sc. 1876, ser. 3, vol. 12, p. 439. The Minerals and Mineral Localities of North Carolina, by Genth and Kerr. *Raleigh*, 1885, p. 14.	12·0
66	1*g*	**Black Mountain**, 15 m. E. of Asheville, Buncombe County, N. Carolina,U.S.A. Found about 1839, and described by Shepard in 1847.	Amer. Jour. Sc. 1847, ser. 2, vol. 4, p. 82.	71·5
67	1*h*	**Asheville** (Baird's Plantation, 6 m. N. of), Buncombe County, N. Carolina, U.S.A. Found loose in the soil : described by Shepard in 1839.	Amer. Jour. Sc. 1839, ser. 1, vol. 36, p. 81; and 1847, ser. 2, vol. 4, p. 79.	114·9
68	1*h*	**Haywood County**, N. Carolina, U.S.A. Date of fiud unknown : described in 1854 by Shepard.	Amer. Jour. Sc. 1854, ser. 2, vol. 17, p. 327.	—
69	1*k*	**Chesterville**, Chester County, S. Carolina, U.S.A. Ploughed up several years before 1849, when it was described by Shepard.	Amer. Jour. Sc. 1849, ser. 2, vol. 7, p. 449.	2,250·4
70	1*k*	**Laurens County**, S. Carolina, U.S.A. Found in 1857 : described by Hidden in 1886.	Amer. Jour. Sc. 1886, ser. 3, vol. 31, p. 463.	63·5
71	1*k*	**Ruff's Mountain**, Lexington County, S. Carolina, U.S.A. Date of find not stated : described by Shepard in 1850.	Amer. Jour. Sc. 1850, ser. 2, vol. 10, p. 128.	498·7
72	1*k*	**Lexington County**, S. Carolina, U.S.A. Found in 1880 : described by Shepard in 1881.	Amer. Jour. Sc. 1881, ser. 3, vol. 21, p. 117.	271·5
73	1*l*	**Union County**, Georgia, U.S.A. Found in 1853 : described by Shepard in 1854.	Amer. Jour. Sc. 1854, ser. 2, vol. 17, p. 328.	55·0
74	1*l*	**Whitfield County** (Dalton), Georgia, U.S.A. First specimen found in 1877 : particulars of find, and description given by Hidden in 1881. A second specimen was found in 1879, and described by Shepard in 1883.	Amer. Jour. Sc. 1881, ser. 3, vol. 21, p. 286. Amer. Jour. Sc. 1883, ser. 3, vol. 26, p. 337.	146·4

No.	Pane.	Name of Meteorite and Place of Find.	Report of Find.	Weight in grams.
75	1*l*	**Losttown** (2½ m. S.W. of), Cherokee County, Georgia, U.S.A. Ploughed up in 1868 : described in the same year by Shepard.	Amer. Jour. Sc. 1868, ser. 2, vol. 46, p. 257.	6·4
76	1*m*	**Canton,** Cherokee County, Georgia, U.S.A. Ploughed up in 1894: described by Howell in 1895.	Amer. Jour. Sc. 1895, ser. 3, vol. 50, p. 252.	335.0
77	1*l*	**Holland's Store,** Chattooga County, Georgia, U.S.A. Found in 1887 : described by Kunz in the same year.	Amer. Jour. Sc. 1887, ser. 3, vol. 34, p. 471.	204·0
78	1*l*	**Putnam County,** Georgia, U.S.A. Found in 1839: described by Willet in 1854.	Amer. Jour. Sc. 1854, ser. 2, vol. 17, p. 331.	112·5
79	1*l*	**Chulafinnee,** Cleberne County, Alabama, U.S.A. Ploughed up in 1873: described by Hidden in 1880.	Amer. Jour. Sc. 1880, ser. 3, vol. 19, p. 370.	60·0
80	1*l*	**Auburn,** Lee County, Alabama, U.S.A. Ploughed up some years before 1869, when it was described by Shepard.	Amer. Jour. Sc. 1869, ser. 2, vol. 47, p. 230.	37·5
81	1*l*	**Summit,** Blount County, Alabama, U.S.A. Known since 1890: described by Kunz in the same year.	Amer. Jour. Sc. 1890, ser. 3, vol. 40, p. 322.	47·7
82	1*l*	**Walker County,** Alabama, U.S.A. Found in 1832, described by Troost in 1845.	Amer. Jour. Sc. 1845, ser. 1, vol. 49, p. 344.	22,295·0
83	1*l*	**Claiborne** (Lime Creek), Clarke County, Alabama, U.S.A. Mentioned in 1834, described by Jackson in 1838.	Amer. Jour. Sc. 1838, ser. 1, vol. 34, p. 332.	65·2
84	1*l*	**Oktibbeha County,** Mississippi, U.S.A. Found in an Indian tumulus : described by Taylor in 1857.	Amer. Jour. Sc. 1857, ser. 2, vol. 24, p. 293.	—
85	1*n*	**Cocke County** (Cosby's Creek), Tennessee, U.S.A. Described in 1840 by Troost: date of find unknown.	Amer. Jour. Sc. 1840, ser. 1, vol. 38, p. 253.	52,325·0

No.	Pane.	Name of Meteorite and Place of Find.	Report of Find.	Weight in grams.
86	1*l*	**Babb's Mill**, Green County, Tennessee, U.S.A. Turned up by a plough : first mentioned in 1842 : described by Troost in 1845.	Amer. Jour. Sc. 1845, ser. 1, vol. 49, p. 342.	2,164·3
87	1*l*	**Tazewell**, Claiborne County, Tennessee, U.S.A. Turned up by a plough in 1853: described by Shepard in 1854.	Amer. Jour. Sc. 1854, ser. 2, vol. 17, p. 325.	336·5
88	1*l*	**Waldron Ridge**, Claiborne County, Tennessee, U.S.A. Known since 1887: described by Kunz in the same year.	Amer. Jour. Sc. 1887, ser. 3, vol. 34, p. 475.	70·0
89	1*m*	**Cleveland**, Bradley County, Tennessee, U.S.A. This mass was acquired in 1867 by Lea, and described by Genth in 1886.	Proc. Ac. Nat. Sc. Philad. 1886, p. 366.	209·0
90	1*m*	**Jackson County**, Tennessee, U.S.A. Date of find unknown : described in 1846 by Troost.	Amer. Jour. Sc. 1846, ser. 2, vol. 2, p. 357.	91·0
91	1*o*	**Carthage**, Smith County, Tennessee, U.S.A. Found about 1844: described in 1846 by Troost.	Amer. Jour. Sc. 1846, ser. 2, vol. 2, p. 356.	24,570·0
92	1*m*	**Caney Fork**, DeKalb County, Tennessee, U.S.A. Turned up by a plough in the same district, near the mouth of the Caney Fork ("Caryfort"), date not mentioned: described by Troost in 1845.	Amer. Jour. Sc. 1845, ser. 1, vol. 49, p. 341.	4·5
93	1*n*	**Smithville**, De Kalb County, Tennessee, U.S.A. Three masses were ploughed up in 1892-3: described by Huntington in 1894.	Proc. Amer. Ac. Arts & Sci. 1894: new series, vol. 21, p. 251.	1745·0
94	1*l*	**Murfreesboro'**, Rutherford County, Tennessee, U.S.A. Found about 1847-8: described in 1848 by Troost.	Amer. Jour. Sc. 1848, ser. 2, vol. 5, p. 351.	2,794·2
95	1*m*	**Coopertown**, Robertson County, Tennessee, U.S.A. Sent to Smith in 1860: described by him in 1861.	Amer. Jour. Sc. 1861, ser. 2, vol. 31, p. 266.	180·0
96	1*m*	**Kenton County** (8 miles south of Independence), Kentucky, U.S.A. Found in 1889: described by Preston in 1892.	Amer. Jour. Sc. 1892, ser. 3, vol. 44, p. 163.	2,520·0

No.	Pane.	Name of Meteorite and Place of Find.	Report of Find.	Weight in grams.
97	1m, 4l	**Lagrange**, Oldham County, Kentucky, U.S.A. Found in 1860 : described by Smith in 1861.	Amer. Jour. Sc. 1861, ser. 2, vol. 31, p. 265.	217·0
98	1m	**Frankfort** (8 miles S.W. of), Franklin County, Kentucky, U.S.A. Found in 1866 : described (1870) by Smith.	Amer. Jour. Sc. 1870, ser. 2, vol. 49, p. 331.	98·0
99	1m, 4l	**Salt River**, about 20 miles below Louisville, Kentucky, U.S.A. Date of find not mentioned : described by Silliman in 1850.	Proc. Amer. Ass. 1851, p. 36.	524·0
100	1m, 4l	**Nelson County**, Kentucky, U.S.A. Turned up by a plough in 1860 : described by Smith in the same year.	Amer. Jour. Sc. 1860, ser. 2, vol. 30, p. 240.	3,907·6
101	1m	**Casey County**, Kentucky, U.S.A. Mentioned in 1877 by Smith.	Amer. Jour. Sc. 1877, ser. 3, vol. 14, p. 246.	45·5
102	1m	**Scottsville**, Allen County, Kentucky, U.S.A. Found in 1867 : described by Whitfield in 1887.	Amer. Jour. Sc. 1887, ser. 3, vol. 33, p. 500.	409·6
103	1n	**Smithland**, Livingston County, Kentucky, U.S.A. Found about 1839–40, and described in 1846 by Troost.	Amer. Jour. Sc. 1846, ser. 2, vol. 2, p. 357.	2,556·2
104	1m	**Marshall County**, Kentucky, U.S.A. Described by Smith in 1860.	Amer. Jour. Sc. 1860, ser. 2, vol. 30, p. 240.	80·3
105	1m	**Wayne County** (near Wooster), Ohio, U.S.A. Found about 1858 : described by Smith in 1864.	Amer. Jour. Sc. 1864, ser. 2, vol. 38, p. 385.	5·2
106	1m	**Grand Rapids**, Kent County, Michigan, U.S.A. Found in 1883 about 3 feet below the surface : reported by Eastman in 1884.	Amer. Jour. Sc. 1884, ser. 3, vol. 28, p. 299.	1,146·0
107	1n	**Howard County** (7 miles S.E. of Kokomo), Indiana, U.S.A. Found in 1862 or 1870 at a depth of 2 feet : described by Cox in 1872 and by Smith in 1874.	Amer. Jour. Sc. 1873, ser. 3, vol. 5, p. 155 ; and 1874, ser. 3, vol. 7, p. 391.	38·0
108	1n	**Plymouth**, Marshall County, Indiana, U.S.A. Found in 1893 by a ploughman : described by Ward, in 1895.	Amer. Jour. Sc. 1895, ser. 3, vol. 49, p. 53.	446·0

No.	Pane.	Name of Meteorite and Place of Find.	Report of Find.	Weight in grams.
109	1m	**Independence County** (about 7 miles east of Batesville), Arkansas, U.S.A. Found in 1884 : described by Hidden in 1886.	School of Mines Quarterly, vol. 7, No. 2, Jan. 1886.	372·0
110	1n	**South-East Missouri,** U.S.A. Found in 1863 in the Museum of St. Louis, labelled "South-East Missouri:" reported by Shepard in 1869.	Amer. Jour. Sc. 1869, ser. 2, vol. 47, p. 233.	102·5
111	1n	**Butler,** Bates County, Missouri, U.S.A. Turned up by a plough : long afterwards came to the knowledge of Broadhead who mentioned it in 1875.	Amer. Jour. Sc. 1875, ser. 3, vol. 10, p. 401.	389·0
112	1n	**Trenton,** Washington County, Wisconsin, U.S.A. Turned up by a plough in 1858 : described by Dörflinger in 1868.	Smithson. Rep. for 1869 : p. 417.	223·0
113	1m	**Hammond Township,** St. Croix County, Wisconsin, U.S.A. Ploughed up in 1884 : described by Fisher in 1887.	Amer. Jour. Sc. 1887, ser. 3, vol. 34, p. 381.	62·0
114	1o	**Dakota,** U.S.A. Described in 1863 by Jackson.	Amer. Jour. Sc. 1863, ser. 2, vol. 36, p. 259.	223·8
115	1o	**Jamestown** (15 or 20 miles south-east of), Stutsman County, N. Dakota, U.S.A. Found in 1885 : described by Huntington in 1891.	Proc. Amer. Ac. Arts & Sci. 1891, vol. 25, p. 229.	1,627·0
116	1n	**Crow Creek,** Laramie County, Wyoming, U.S.A. Found in 1887 : described by Kunz in 1888.	Amer. Jour. Sc. 1888, ser. 3, vol. 36, p. 276.	583·0
117	1o	**Nebraska** (25 m. N.W. of Fort St. Pierre), U.S.A. Brought away in 1857 : described by Holmes in 1860.	Trans. of St. Louis Acad. of Sc. 1857–60, vol. 1, p. 711.	2,016·0
118	1n	**Russel Gulch,** Gilpin County, Colorado, U.S.A. Found in 1863 : described in 1866 by Smith.	Amer. Jour. Sc. 1866, ser. 2, vol. 42, p. 218.	245·4
119	1n	**Bear Creek,** Denver, Colorado, U.S.A. Found in 1866 : described by Shepard in the same year.	Amer. Jour. Sc. 1866, ser. 2, vol. 42, pp. 250, 286.	52·3
120	1n	**Oroville,** Butte County, California, U.S.A. Found in 1893.		508·0

E

No.	Pane.	Name of Meteorite and Place of Find.	Report of Find.	Weight in grams.
121	1*o*	**Shingle Springs**, El Dorado County, California, U.S.A. Found 1869–70: described by Silliman in 1873.	Amer. Jour. Sc. 1873, ser. 3, vol. 6, p. 18.	84·5
122	1*n*	**Ivanpah**, San Bernardino County, California, U.S.A. Described by Shepard in 1880, shortly after its discovery.	Amer. Jour. Sc. 1880, ser. 3, vol. 19, p. 381.	33·0
123	Sep. Stand, 1*h*	**Cañon Diablo**, Arizona, U.S.A. Found in 1891: described by Foote in the same year.	Amer. Jour. Sc. 1891, ser. 3, vol. 42, p. 413.	82,180·0
124	1*n*	**Costilla Peak**, Cimarron Range, New Mexico, U.S.A. Found in 1881 by a sheep-herder: described by Hills in 1895.	Proc. Colorado Scient. Soc. 1895, p. 1.	1,595·0
125	1*o*	**Capitan Range**, New Mexico, U.S.A. Found in 1893 by a sheep-herder: described by Howell in 1895.	Amer. Jour. Sc. 1895, ser. 3, vol. 50, p. 253.	956·0
126a	1*n*	**Glorieta Mountain**, 1 m. N.E. of Canoncito, Santa Fé County, New Mexico, U.S.A. Found in 1884: described by Kunz in 1885.	Amer. Jour. Sc. 1885, ser. 3, vol. 30, p. 235.	1,527·0
126b	1*n*	A specimen probably from this locality was sent in 1884 to Denver from Albuquerque, New Mexico, as silver bullion: described by Pearce and Eakins in 1884–5.	Proc. Colorado Scient. Soc. 1884, vol. 1, p. 110 ; 1885, vol. 2, pp. 14, 35.	61·3
127	1*n*	**Brazos** River, Wichita County, Texas, U.S.A. Known to the Comanches for many years: removed in 1836: described by Shumard in 1860, and by Mallet in 1884.	Trans. of St. Louis Acad. of Sc. 1857–60, vol. 1, p. 622. Amer. Jour. Sc. 1884, ser. 3, vol. 28, p. 285.	1,395·4
128	1*n*	**Denton County**, Texas, U.S.A. After discovery it remained with a blacksmith for several months ; in 1859 it came into the possession of Shumard, by whom it was described in the following year.	Trans. of St. Louis Acad. of Sc. 1857–60, vol. 1, p. 623.	122·0
129	1*n*	**Red River** (Cross Timbers), Johnson County, Texas, U.S.A. Mentioned in 1808 to Captain Glass, and reported by Gibbs in 1814.	Amer. Min. Jour. by Bruce: 1814, vol. 1, pp. 124, 218. Amer. Jour. Sc. 1824, ser. 1, vol. 8, p. 218.	424·5

No.	Pane.	Name of Meteorite and Place of Find.	Report of Find.	Weight in grams.
130	1*m*	**Carlton,** Hamilton County, Texas,U.S.A. Ploughed up in 1888: described by Howell in 1890.	Proc. Rochester Ac. of Sc., 1890, vol. 1, p. 87.	6,185·0
131	2*l*	**Kendall County,** Texas, U.S.A. Found before 1887.	Verhand. d. Ges. deut. Naturf. u. Ärzte: Theil II., Hälfte I.: p. 166. (Naturw. Abtheil.) 1894.	556·0
132	1*o*	**Fort Duncan,** Maverick County, Texas, U.S.A. Found in 1882: described by Hidden in 1886: similar to Coahuila; perhaps transported from the same district by way of Santa Rosa.	Mineralog. Magaz. 1890,vol. 9, p. 116.	4,520·0
133a	1*o*	**Coahuila,** Mexico. Since 1837 many masses have been brought to Santa Rosa, from a district of small area about 90 miles north-west of that town. An account of a visit by Hamilton was published by Shepard in 1866; he designated the iron by the name Bonanza: eight large masses were removed to the United States by Butcher in 1868.	Mineralog. Maga- zine, 1890, vol. 9, p. 107. .	253,645·8
133b	1*o*	**Sanchez Estate,** Coahuila, Mexico. Found in 1853 by Couch in use as an anvil at Saltillo. It was said to have been brought to that town from the "Saucha Estate," but had probably been acquired still earlier at Santa Rosa, and been got at the north-west locality.	Mineralog. Maga- zine, 1890, vol. 9, p. 113.	573·0
134	1*o*	**Sierra Blanca,** Huejuquilla or Jimenez, Chihuahua, Mexico. The occurrence at Sierra Blanca was recorded in 1784: the only specimen known—that from the Bergemann collec- tion—is now thought to be of doubtful authenticity; in its etched figures it is like Toluca.	Mineralog. Maga- zine, 1890, vol. 9, p. 140.	47·3
135	1*o*	**Concepcion Mass,** Huejuquilla or Jimenez, Chihuahua, Mexico. Masses of iron, some of them probably belonging to one fall, have been known for centuries to exist near Huejuquilla: the Concepcion mass is said to have been trans- ported from Sierra de las Adargas in 1780.	Ann. d.k.k. Naturh. Hofmus. Wien, 1896,vol. 10, p. 274.	15·3

No.	Pane.	Name of Meteorite and Place of Find.	Report of Find.	Weight in grams.
136	1a	**Rancho de la Pila,** Labor de Guadalupe, Durango, Mexico. Ploughed up in 1882 : described by Häpke in 1883.	Mineralog. Magazine, 1890, vol. 9, p. 153.	46,512·4
137	2b	**San Francisco del Mezquital,** Durango, Mexico. Brought from Mexico by General Castelnau, and described in 1868 by Daubrée. The above is the old name for the capital of Mezquital.	Mineralog. Magazine 1890, vol. 9, p. 154.	7,120·0
138	1o	**Bella Roca,** Sierra de San Francisco, Santiago Papasquiaro, Durango, Mexico. Acquired by Ward in 1888 : described by Whitfield in 1889.	Amer. Jour. Sci. 1889, ser. 3, vol. 37, p. 439.	3,542·0
139	1o	**Descubridora,** Catorce, San Luis Potosi, Mexico. Found before 1780, and described by a committee in 1872.	Mineralog. Magazine, 1890, vol. 9, p. 157.	29·5
140	4l	**Charcas,** San Luis Potosi, Mexico. Mentioned in 1804 by Sonneschmid; it was then at the corner of the church, and was said to have been brought from San José del Sitio, 12 leagues distant. In 1866 it was removed to Paris.	Mineralog. Magazine, 1890, vol. 9, p. 160.	332·3
141	2c,4l	**Zacatecas,** Mexico. Mentioned in 1792; it was said to have been found long before near the Quebradilla Mine.	Mineralog. Magazine, 1890, vol. 9, p. 162.	3,846·9
142	1a 2c 4l	**Toluca** Valley, Mexico. Before 1776 it was known that masses of iron occurred in the neighbourhood of Xiquipilco, Valley of Toluca.	Mineralog. Magazine, 1890, vol. 9, p. 164.	106,547·7
143	1o	**Yanhuitlan,** Misteca alta, Oaxaca, Mexico. Mentioned by Del Rio in 1804.	Mineralog. Magazine, 1890, vol. 9, p. 171.	316·5
144	Dr.	**Lucky Hill,** St. Elizabeth, Jamaica. Found in 1885 about two feet below the surface.		
145a	2c	**Santa Rosa** (Tocavita), near Tunja, Boyaca River, New Granada, S. America. In 1824 Rivero and Boussingault made known a large mass of iron in use as an anvil at Santa Rosa : with other small pieces it had been found on a neighbouring hill, called Tocavita, in 1810 : they collected several specimens themselves.	Ann. Chim. Phys. 1824, vol. 25, p. 438.	101·0

No.	Pane.	Name of Meteorite and Place of Find.	Report of Find.	Weight in grams.
145b	2c	**Rasgata,** New Granada, S. America. Other masses of iron were seen by Rivero and Boussingault at Rasgata, and were said to have been found there. From the similarity of their characters it is probable that Santa Rosa and Rasgata fell at the same time.	Ann. Chim. Phys. 1824, vol. 25, p. 442.	58·5
146	2d	**Tarapaca Desert** (46 miles from Hemalga), Chili. Found in 1840: described by Greg in 1855.	Phil. Mag. 1855, ser. 4, vol. 10, p. 12.	1,655·8
147	2a	**Mount Hicks,** Mantos Blancos, about 40 miles from Antofagasta, Atacama, Chili. Found about 1876, and described by L. F. in 1889.	Mineralog. Magazine, 1889, vol. 8, p. 257.	9,015·0
148	2d	**Serrania de Varas,** Atacama, Chili. Found about 1875, and described by L. F. in 1889.	Mineralog. Magazine, 1889, vol. 8, p. 258.	1,168·0
149	2d	**Cachiyuyal,** Atacama, Chili. Found in 1874: described by Domeyko in 1875.	Mineralog. Magazine, 1889, vol. 8, p. 259.	28·0
150	2d	**Ilimaë,** Atacama, Chili. Known since 1870: described by Tschermak in 1872.	Mineralog. Magazine, 1889, vol. 8, p. 260.	39·4
151	2h	**Merceditas,** 10 or 12 leagues East of Chañaral, Atacama, Chili. Known since 1884: described by Howell in 1890.	Proc. Rochester Ac. of Sc. 1890, vol. 1, p. 99.	1,917·0
152	Below 1a	**Pan de Azucar,** Atacama, Chili. Found about 67 miles from the port of Pan de Azucar in 1887.		20,250.0
153	2d	**Juncal,** Atacama, Chili. Found in 1866 between Rio Juncal and the Salinas de Pedernal: had possibly been transported to that place: described by Daubrée in 1868.	Mineralog. Magazine, 1889, vol. 8, p. 261.	75·0
154	2d	**Puquios,** Copiapo, Atacama, Chili. Found about 1885: described by Howell in 1890.	Proc. Rochester Ac. of Sc. 1890, vol. 1, p. 89.	176·0
155	2d	**The Joel Iron,** Atacama, Chili. Found in 1858 in an unspecified part of the desert: described by L. F. in 1889.	Mineralog. Magazine, 1889, vol. 8, p. 263.	1, 144·0

No.	Pane.	Name of Meteorite and Place of Find.	Report of Find.	Weight in grams.
156	2*d*	**Barranca Blanca,** between Copiapo and Catamarca, South America. Found in 1855, and described by L. F. in 1889.	Mineralog. Magazine, 1889, vol. 8, p. 262.	11,915·0
157	2*d*	**Chili.** Owing to an interchange of labels, the specimen was described in 1868 by Daubrée as having been found in an unspecified locality in Chili. According to Domeyko it was supposed to have been found in the Cordillera de la Dehesa, near Santiago.	Mineralog. Magazine, 1889, vol. 8, p. 256.	2·0
158	Sep. Stand, 4*c*	**Otumpa,** Gran Chaco Gualamba, Argentine Republic. The occurrence of metallic iron at this locality having been reported, Don Rubin de Celis was sent in 1783 to investigate the matter : his report was published in 1788.	Phil. Trans. 1788, vol. 78, pp. 37, 183. Mineralog. Magaz. 1889, vol. 8, p. 229.	637,000·0
159	2*d*	**Bendegó River,** Bahia, Brazil. Found in 1784 : described by Mornay in 1816.	Phil. Trans. 1816, vol. 106, p. 270.	3,115·0
160	2*d*	**Santa Catharina** (Morro do Rocio), Rio San Francisco do Sul, Brazil. Discovered in 1875 : described by Lunay in 1877 : it is regarded by some mineralogists as probably of terrestrial origin.	Comptes Rendus, 1877, vol. 85, p. 84.	6,399·0
161	2*d*	**Locality unknown** (from Prof. Wöhler's Collection). Described by Wöhler in 1852.	Ann. Chem. Pharm. 1852, vol. 81, p. 253.	30·5
162	2*d*	**Locality unknown** (from Smithsonian Museum Collection). Described by Shepard in 1881.	Amer. Jour. Sc. 1881, ser. 3, vol. 22, p. 119.	5·5
163	2*d*	**Locality unknown** (from United States National Museum Collection). Slice of a complete meteorite which was found in a collection of minerals formed by the late Col. J. J. Abert : described by Riggs in 1887.	Amer. Jour. Sc. 1887, ser. 3, vol. 34, p. 59.	47·0

II. SIDEROLITES

(consisting chiefly of nickeliferous iron and silicates, both in large proportion).

A. FALL RECORDED.

[Arranged chronologically.]

No.	Pane.	Name of Meteorite and Place of Fall.	Date of Fall.	Weight in grams.
164	2e	**Taney County**, Missouri, U.S.A. . . . A fragment, sent from Taney County, Missouri, about 1857–8, was described by Shepard in 1860. *Amer. Jour. Sc.* 1860, ser. 2, vol. 30, p. 205. A fragment of a meteorite was given to Cox by Judge Green of Crawford County: no mention of place or date of find. *Sec. Rep. Geol. Reconn. Arkansas,* 1860, p. 408. Green's fragment was described under the name of Newton County (Arkansas) by Smith in 1865. *Amer. Jour. Sc.* 1865, ser. 2, vol. 40, p. 213. A large mass was obtained by Kunz and reported by him in 1887 to have really fallen in Taney County, Missouri, about thirty years before, and to have been afterwards taken to Newton County, Arkansas. *Amer. Jour. Sc.* 1887, ser. 3, vol. 34, p. 467.	Fell about 1857–8.	2404·5
165	2e	**Lodran**, Mooltan, Punjaub, India . .	Oct. 1, 1868.	66·5
166	2a	**Estherville**, Emmet County, Iowa, U.S.A.	May 10, 1879.	116,903·0
167	2e	**Veramin**, Teheran, Persia	Fell 1879–80.	53·85

Siderolites.

B. Fall not Recorded.

[Arranged geographically.]

No.	Pane.	Name of Meteorite and Place of Find.	Report of Find.	Weight in grams.
168	2c	**Hainholz**, Minden, Westphalia. Found in 1856: described by Wöhler in 1857.	Pogg. Ann. 1857, vol. 100, p. 342.	484·1
169a	2c	**Steinbach**, Erzgebirge, Saxony. Reported as "native iron" by J. G. Lehmann in 1751.	Kurze Einleitung in einige Theile der Bergwerks-Wissen-schaft, 1751, p. 79.	132·0
169b	2d	**Rittersgrün**, Erzgebirge, Saxony. Found in 1847: reported by Breithaupt in 1861. According to Weisbach it was really fonnd in 1833.	Zeitsch. deutsch. geol. Gesell. 1861, vol. 13 p. 148. Der Eisenmeteorit von Rittersgrün im sächsischen Erzge-birge: von A. W. : Freiberg, 1876.	694·2
169c	2c	**Breitenbach**, Erzgebirge, Bohemia. Found in 1861: described by Maskelyne in 1871. Steinbach, Rittersgrün, and Breitenbach are within five English miles of each other, on the border of Saxony and Bohemia; the siderolites probably fell at the same time. Breithaupt suggests that this was the fall reported to have taken place at Whitsuntide in the year 1164: Buchner (p. 124) suggests a fall which took place between 1540 and 1550.	Phil. Trans. 1871, vol. 161, p. 359. Berg-und hütt.Zei-tung, 1862, Jahrg. 21, p. 321.	6,231·0
170	2c	**Brahin**, Minsk, Russia. Found in 1809, 1810 or 1820.	Bull. des. Sc. par la Soc.philom., *Paris*, 1823, p. 86. Partsch's Die Me-teoriten zu Wien. 1843, p. 90. Erman's Archiv. f. wiss. Kunde von Russland, 1846, vol. 5, p. 183.	22·2
171	2c,4c	**The Pallas iron.** Found in 1749 between the Ubei and Sisim rivers, Jeniseisk, Asiatic Russia : re-ported by Pallas in 1776.	Reise d. versch. Prov. d. russ. Reichs : von P. S. Pallas. St. Peters-burg,1776. Part iii. p. 411.	3,735·8

No.	Pane.	Name of Meteorite and Place of Find.	Report of Find.	Weight in grams.
172	2e	**Pavlodar,** Semipalatinsk, Asiatic Russia. Found in 1885.		56·3
173	2e	**River Senegal,** West Africa. "Native Iron" was found by Compagnon in 1716 to be in very common use in many parts of the kingdoms of Bambuk and Siratik.	Allgemeine Historie der Reisen zu Wasser und Lande : von J. J. Schwabe. Leipzig, 1748, vol. 2, Book 5, Ch. 13, p. 510.	396·0
174	2e	**Powder Mill Creek,** Cumberland County, Tennessee, U.S.A. Found in 1887: described in the same year by Whitfield and Kunz.	Amer. Jour. Sc. 1887, ser. 3, vol. 34, pp. 387, 476.	1148·0
175	2f	**Eagle Station,** Carroll County, Kentucky, U.S.A. Found in 1880, and described by Kunz in 1887.	Amer. Jour. Sc. 1887, ser. 3, vol. 33, p. 228.	708·0
176	2f	**Brenham Township,** Kiowa County, Kansas, U.S.A. Found about 1886: described by Kunz in 1890.	Amer. Jour. Sc. 1890, ser. 3, vol. 40, p. 312.	2,011·0
177	2f	**Tucson,** Arizona, U.S.A. Two large masses, long preserved at Tucson, had been transported to that town from the Puerto de los Muchachos, a pass about 20 or 30 miles south of Tucson. Their existence has been known for centuries. One of them has been termed the Signet or Irwin-Ainsa iron, the other the Carleton iron.	Mineralog. Magazine, 1890, vol. 9, p. 16.	161·0 262·0
178	Sep. Stand, 2f	**Imilac,** Atacama, Chili. Known in 1822: probably the specimen found at Campo de Pucará had been carried from Imilac.	Mineralog. Magazine, 1889, vol. 8, p. 243.	227,328·0
179	2e	**Vaca Muerta,** Atacama, Chili. Mentioned in 1861, and described in 1864 by Domeyko as found at Sierra de Chaco. Specimens probably got from the same place are known by various names (Mejillones, Jarquera or Janacera Pass, &c.)	Mineralog. Magazine, 1889, vol. 8, p. 234.	7,283·0

No.	Pane.	Name of Meteorite and Place of Find.	Report of Find.	Weight in grams.
180	2e	**Llano del Inca,** 35 leagues S.E. of Taltal, Atacama, Chili.	Proc. Rochester Ac. of Sci. 1890, vol. 1, p. 93.	376·0
181	2e	**Doña Inez,** Atacama, Chili. The meteorites of Llano del Inca and Doña Inez were found in these localities in 1888, and were described by Howell in 1890 : "polished sections of the two meteorites are in many cases not distinguishable," and Howell is inclined to think that they belong to a single fall. (Some of the polished faces are not to be distinguished from those of Vaca Muerta.) L. F.		1,016·0
182	2e	**Copiapo,** Chili. Numerous masses of this type have been brought to Copiapo since 1863 : some of them, owing to an interchange of labels, have been supposed to come from the Sierra de la Dehesa (Deesa), near Santiago.	Mineralog. Magazine, 1889, vol. 8, p. 255.	769·5

Aerolites. A. Fall recorded.

75

III. AEROLITES

or Meteoric Stones

(consisting generally of one or more silicates, interspersed with isolated particles of nickeliferous iron, troilite, &c.).

A. FALL RECORDED.

[Arranged chronologically.]

No.	Pane.	Name of Meteorite and Place of Fall.	Date of Fall.	Weight in grams.
183	4c	**Ensisheim**, Elsass, Germany . .	Nov. 16, 1492	458·0
184	2g	**Schellin**, near Stargard, Pomerania, Prussia.	April 11, 1715	—
185	2g	**Plescowitz**, near Reichstadt, Bohemia .	June 22, 1723	25·6
186	4c	**Ogi**, Hizen, Kiusiu, Japan . . .	Fell about 1730	4,185·0
187	4c	**Tabor** (Plan, Strkow), Bohemia . .	July 3, 1753	151·0
188	2g	**Luponnas**, Ain, France . . .	Sept. 7, 1753	7·7
189	2g	**Albareto**, Modena, Italy . . .	July 1766	53·0
190	4c	**Lucé** (Maine), Sarthe, France .	Sept. 13, 1768	11·9
191	2g	**Mauerkirchen**, Upper Austria . .	Nov. 20, 1768	302·0
192	2g	**Eichstädt**, Bavaria	Feb. 19, 1785	13·8
193	2h	**Kharkov** (Bobrik), Russia . . .	Oct. 12, 1787	437·2
194	2h	**Barbotan**: (a) Barbotan, (b) Roquefort, } Landes, France.	July 24, 1790	{ 712·5 145·5
195	4c	**Siena**, Cosona, Italy	June 16, 1794	128·7
196	4b	**Wold Cottage**, Thwing, Yorkshire .	Dec. 13, 1795	20,111·0
197	2g	**Bjelaja Zerkov**, Kiev, Russia . .	Jan. 15 or 16, 1796	9·2
198	2g	**Salles**, near Villefranche, Rhône, France.	March 12, 1798	165·0
199	4c	**Krakhut**, Benares, India . . .	Dec. 19, 1798	510·6
200	2k, 4c	**L'Aigle**, Orne, France. . .	April 26, 1803	2,242·0
201	2h	**Apt** (Saurette), Vaucluse, France .	Oct. 8, 1803	37·4
202	3n	**Mässing** (St. Nicholas), Bavaria . .	Dec. 13, 1803	—
203	2g	**Darmstadt**, Hesse, Germany . .	Fell before 1804	1·6
204	4d	**High Possil**, near Glasgow, Scotland .	April 5, 1804	91·3
205	2g	**Hacienda de Bocas**, San Luis Potosí, Mexico.	Nov. 24, 1804	—
206	2g	**Doroninsk**, Irkutsk, Asiatic Russia .	April 6, 1805	8·9
207	2g	**Asco**, Corsica	Nov. 1805	—
208	4n	**Alais**, Gard, France	March 15, 1806	13·0
209	2h	**Timochin**, Juchnov, Smolensk, Russia.	March 25, 1807	138·5
210	2k, 4c	**Weston**, Connecticut, U.S.A. . .	Dec. 14, 1807	1,034·5
211	2g	**Cusignano**, Noceto, Parma, Italy .	April 19, 1808	9·7
212	3n 4d 4c	**Stannern**: (a) Stannern, (b) Langenpiernitz, } Iglau, Moravia, Austria.	May 22, 1808	{ 1,570·0 13·8

No.	Pane.	Name of Meteorite and Place of Fall.	Date of Fall.	Weight in grams.
213	2h	**Lissa**, Bunzlau, Bohemia . . .	Sept. 3, 1808	169·6
214	2g	**Moradabad**, North-West Provinces, India.	Fell in 1808	17·1
215	2h	**Kikino**, Viasma, Smolensk, Russia .	Fell in 1809	25·0
216	2h	**Mooresfort**, County Tipperary, Ireland.	Aug. 1810	243·4
217	2h	**Charsonville**: (a) Charsonville, (b) Bois de Fontaine, (c) Fragment of a stone labelled *Chartres*. Meung, Loiret, France.	Nov. 23, 1810	{ 108·6 / 2,227·0 / 20·0 }
218	2h	**Kuleschovka**, Poltava, Russia . .	March 12, 1811	57·9
219	2h	**Berlanguillas**, near Burgos, Spain .	July 8, 1811	26·5
220	2h	**Toulouse** (Grenade), Haute Garonne, France.	April 10, 1812	31·9
221	2h	**Erxleben**, Magdeburg, Prussia . .	April 15, 1812	31·5
222	2l,4o	**Chantonnay**, Vendée, France . .	Aug. 5, 1812	1,352·3
223	2k	**Limerick** (Adare, Faha, &c.), Ireland .	Sept. 10, 1813	114·5
224	3n	**Luotolax**, Wiborg, Finland . .	Dec. 13, 1813	20·7
225	2h	**Gurram Konda**, between Punganur and Kadapa, Madras, India.	Fell in 1814	9·8
226	2k	**Bachmut**, Ekaterinoslav, Russia . .	Feb. 15, 1814	40·8
227	2k	**Agen**, Lot-et-Garonne, France . .	Sept. 5, 1814	40·6
228	2k	**Chail**, Allahabad, North-West Provinces, India.	Nov. 5, 1814	—
229	2l	**Durala**, N.W. of Kurnal, Punjaub, India	Feb. 18, 1815	12,588·9
230	4o	**Chassigny**, Haute Marne, France .	Oct. 3, 1815	41·3
231	2k	**Zaborzika**, Czartorya, Volhynia, Russia	April 11, 1818	9·2
232	4n	**Seres**, Macedonia, Turkey . .	June 1818	399·6
233	2l	**Slobodka**, Juchnov, Smolensk, Russia .	Aug. 10, 1818	27·5
234	3n	**Jonzac**, Charente Inférieure, France .	June 13, 1819	9·0
235	2l	**Pohlitz**, near Gera, Reuss, Germany .	Oct. 13, 1819	86·9
236	2l	**Lixna**, Dünaburg, Vitebsk, Russia .	July 12, 1820	'59·5
237	4o	**Juvinas**, near Libonnez, Ardèche, France	June 15, 1821	940·0
238	2l	**Angers**, Maine-et-Loire, France . .	June 3, 1822	22·3
239	2l	**Agra** (Kadonah), India . . .	Aug. 7, 1822	38·8
240	2l	**Epinal** (La Baffe), Vosges, France .	Sept. 13, 1822	1·6
241	2l,4h	**Futtehpur**: (a) Futtehpur, (b) Bithur N.West Provinces, India	Nov. 30, 1822	{ 1,286·0 / 136·0 }
242	2l	**Umballa** (40 miles S.W. of), Punjaub, India.	Fell in 1822–3	20·6
243	3n	**Nobleborough**, Lincoln County, Maine, U.S.A.	Aug. 7, 1823	—
244	3m	**Renazzo**, Cento, Ferrara, Italy . .	Jan. 15, 1824	15·0
245	2m	**Zebrak**, near Horowitz, Bohemia . .	Oct. 14, 1824	83·9
246	2m	**Nanjemoy**, Charles County, Maryland, U.S.A.	Feb. 10, 1825	325·5
247	2l	**Honolulu**, Hawaii, Sandwich Islands .	Sept. 27, 1825	81·0
248	2m	**Pavlograd**, Ekaterinoslav, Russia .	May 19, 1826	160·8

No.	Pane.	Name of Meteorite and Place of Fall.	Date of Fall.	Weight in grams.
249	2m	Mhow, Azamgarh District, North-West Provinces, India.	Feb. 16, 1827	163·5
250	2m	Drake Creek, Nashville, Tennessee, U.S.A.	May 9, 1827	19·4
251	3n	Bialystock (Jasly), Grodno, Russia .	Oct. 5, 1827	3·7
252	2m	Richmond, Henrico County, Virginia, U.S.A.	June 4, 1828	169·5
253	2m	Forsyth, Georgia, U.S.A. . .	May 8, 1829	72·5
254	2m	Deal, near Long Branch, New Jersey, U.S.A.	Aug. 14, 1829	—
255	2m	Krasnoi-Ugol, Rjäsan, Russia . .	Sept. 9, 1829	—
256	2m	Launton, Bicester, Oxfordshire . .	Feb. 15, 1830	1,022·0
257	2m	Perth (N. Inch of), Scotland . .	May 17, 1830	1·5
258	2m	Vouillé, near Poitiers, Vienne, France .	May 13, 1831	60·9
259	2m	Wessely, Hradisch, Moravia, Austria .	Sept. 9, 1831	3·1
260	2m	Blansko, Brünn, Moravia, Austria .	Nov. 25, 1833	—
261	2m	Okniny, Kremenetz, Volhynia, Russia .	Dec. 27, 1833	7·0
262	2m	Charwallas, near Hissar, Delhi, India.	June 12, 1834	37·8
263	2m	Mascombes, Corrèze, France . .	Jan. 31, 1835	5·0
264	2n	Aldsworth, near Cirencester, Gloucestershire.	Aug. 4, 1835	525·4
265	3n	Aubres, Nyons, Drôme, France . .	Sept. 14, 1836	488·0
266	2n	Macao, Rio Grande do Norte, Brazil .	Nov. 11, 1836	6·4
267	2n	Nagy-Diwina, near Budetin, Trentschin, Hungary.	July 24, 1837	3·0
268	2n	Esnandes, Charente Inférieure, France.	Aug. 1837	3·0
269	2m	Kaee, Sandee District, Oude, India .	Jan. 29, 1838	209·2
270	2n	Akburpur, Saharanpur, North-West Provinces, India.	April 18, 1838	1,568·7
271	2n	Chandakapur, Berar, India . .	June 6, 1838	760·7
272	2m	Montlivault, Loir-et-Cher, France .	July 22, 1838	11·0
273	3m,4n	Cold Bokkeveldt, Cape Colony .	Oct. 13, 1838	1,057·0
274	2n	Little Piney, Pulaski County, Missouri, U.S.A.	Feb. 13, 1839	103·9
275	2n	Karakol, Ajagus, Kirghiz Steppes, Russia.	May 9, 1840	2·0
276	2n	Uden, North Brabant, Netherlands .	June 12, 1840	5·5
277	2m	Cereseto, near Ottiglio, Alessandria, Piedmont, Italy.	July 17, 1840	124·2
278	2n	Grüneberg, Heinrichsau, Prussian Silesia	March 22, 1841	30·8
279	2m	Château-Renard, Triguères, Loiret, France.	June 12, 1841	3,290·0
280	2n	Milena, Warasdin, Croatia, Austria .	April 26, 1842	25·4
281	2n	Aumières, Lozère, France . . .	June 4, 1842	43·0
282	4o	Bishopville, Sumter County, S. Carolina, U.S.A.	March 25, 1843	512·0
283	2n,4n	Utrecht (Blaauw-Kapel), Netherlands .	June 2, 1843	69·8
284	3n	Manegaum, near Eidulabad, border of Khandeish, India.	June 29, 1843	11·4

No.	Pane.	Name of Meteorite and Place of Fall.	Date of Fall.	Weight · in grams.
285	2o	**Klein-Wenden,** near Nordhausen, Erfurt, Prussia.	Sept. 16, 1843	5·5
286	2n	**Cerro Cosina,** near Dolores Hidalgo, San Miguel, Guanaxuato, Mexico.	Jan. 1844	42·1
287	2n	**Killeter,** County Tyrone, Ireland . .	April 29, 1844	104·7
288	2o	**Favars,** Aveyron, France . . .	Oct. 21, 1844	6·0
289	3n	**Le Teilleul,** Manche, France . .	July 14, 1845	1·9
290	2o	**Monte Milone** (now called Pollenza), Macerata, Italy.	May 8, 1846	8·1
291	2o	**Cape Girardeau,** Missouri, U.S.A.	Aug. 14, 1846	78·7
292	2n	**Schönenberg,** Mindelthal, Schwaben, Bavaria.	Dec. 25, 1846	42·0
293	2o	**Linn County** (Hartford), Iowa, U.S.A.	Feb. 25, 1847	942·5
294	2o	**Castine,** Hancock County, Maine, U.S.A.	May 20, 1848	2·7
295	3n	**Marmande,** Aveyron, France .	July 4, 1848	4·9
296	2n	**Ski,** Amt Akershuus, Norway . .	Dec. 27, 1848	5·6
297	2o	**Cabarras County,** N. Carolina, U.S.A.	Oct. 31, 1849	385·5
298	2o	**Kesen,** Japan.	June 13, 1850	1,281·0
299	3n	**Shalka,** Bancoorah, Bengal, India. .	Nov. 30, 1850	1,132·0
300	2n	**Gütersloh,** Westphalia, Prussia . .	April 17, 1851	109·2
301	2o	**Quinçay,** Vienne, France . .	Summer, 1851	10·0
302	2o	**Nulles,** Catalonia, Spain . . .	Nov. 5, 1851	4·5
303	4p	**Nellore** (Yatoor), Madras, India . .	Jan. 23, 1852	11,287·0
304	2o, 1d	**Mezö-Madaras,** Transylvania . .	Sept. 4, 1852	733·7
305	2o	**Borkut,** Marmoros, Hungary . .	Oct. 13, 1852	40·0
306	4o	**Bustee,** between Goruckpur and Fyzabad, India.	Dec. 2, 1852	1,000·0
307	2o	**Girgenti,** Sicily	Feb. 10, 1853	233·5
308	3c	**Segowlie,** Bengal, India . . .	March 6, 1853	1,205·7
309	2o	**Duruma,** Wanikaland, E. Africa . .	Fell in 1853	1·2
310	2o	**Oesel** (Gesinde Kaande, near Piddul), Baltic Sea.	May 11, 1855	17·9
311	3c	**Gnarrenburg** (Bremervörde), Hanover	May 13, 1855	808·0
312	3c	**St. Denis - Westrem,** near Ghent, Belgium.	June 7, 1855	1·3
313	4o	**Petersburg,** Lincoln County, Tennessee, U.S.A.	Aug. 5, 1855	52·8
314	3c	**Trenzano,** Brescia, Italy . . .	Nov. 12, 1856	157·8
315	3c, 3a	**Parnallee,** Madras, India . . .	Feb. 28, 1857	61,361·0
316	3c	**Heredia,** San José, Costa Rica . .	April 1, 1857	54·0
317	3c	**Stavropol,** north side of the Caucasus, Russia.	April 5, 1857	22·6
318	3m	**Kaba,** Debreczin, Hungary . . .	April 15, 1857	104·2
319	3c	**Les Ormes,** near Joigny, Yonne, France	Oct. 1, 1857	12·2
320	3c	**Ohaba,** near Karlsburg, Transylvania .	Oct. 11, 1857	39·6
321	4n	**Pegu** (Quenggouk), British Burmah .	Dec. 27, 1857	654·0
322	3c	**Kakowa,** Temeser Banat, Hungary .	May 19, 1858	160·6
323	3d	**Ausson:** (a) Ausson,} Haute Garonne, (b) Clarac, } France.	Dec. 9, 1858	{ 367·2 { 110·3
324	3c	**Molina,** Murcia, Spain . . .	Dec. 24, 1858	6·1

No.	Pane.	Name of Meteorite and Place of Fall.	Date of Fall.	Weight in grams.
325	3d	**Harrison County**, Indiana, U.S.A. .	March 28, 1859	38·7
326	3d	**Beuste**, near Pau, Basses-Pyrénées, France.	May 1859	40·5
327	3d	**Bethlehem**, near Albany, New York, U.S.A.	Aug. 11, 1859	—
328	3d	**Pampanga** (Mexico), Philippine Islands	Fell in 1859	1·8
329	3d	**Alessandria** (San Giuliano Vecchio), Piedmont, Italy.	Feb. 2, 1860	35·0
330	4n	**Khiragurh**, S.E. of Bhurtpur, India .	March 28, 1860	353·3
331	2o, 3b	**New Concord**, Muskingum County, Ohio, U.S.A.	May 1, 1860	19,519·0
332	3d	**Kusiali**, Kumaon, India . . .	June 16, 1860	4·1
333	2n	**Dhurmsala**, Kangra, Punjaub, India .	July 14, 1860	12,407·0
334	4h	**Butsura** { (Qutahar Bazaar) (Chireya) (Piprassi) (Bulloah) } Bengal, India.	May 12, 1861	{ 13,071·5 843·0 5,060·0 158·5
335	3d	**Canellas**, near Barcelona, Spain . .	May 14, 1861	1·5
336	3m	**Grosnaja**, Banks of the Terek, Caucasus, Russia.	June 28, 1861	160·0
337	2o	**Klein-Menow**, Alt-Strelitz, Mecklenburg, Germany	Oct. 7, 1862	1,132·0
338	3d	**Pulsora**, N.E. of Rutlam, Indore, Central India.	March 16, 1863	48·0
339	3d	**Buschhof**, Courland, Russia . .	June 2, 1863	98·1
340	3d	**Pillistfer** (Aukoma), Livland, Russia .	Aug. 8, 1863	157·2
341	3d	**Shytal**, 40 miles north of Dacca, India .	Aug. 11, 1863	462·7
342	3d	**Tourinnes-la-Grosse**, Tirlemont, Belgium.	Dec. 7, 1863	203·1
343	3d	**Manbhoom**, Bengal, India. . .	Dec. 22, 1863	122·9
344	3d	**Nerft**, Courland, Russia . . .	April 12, 1864	69·5
345	3m,4d	**Orgueil**, near Montauban, Tarn-et-Garonne, France.	May 14, 1864	621·4
346	3e	**Dolgovoli**, Volhynia, Russia . .	June 26, 1864	1·5
347		**Supuhee :** } Goruckpur District, India.		{ 4,050·6
	2o	(a) Mouza Khoorna, Sidowra,	Jan. 19, 1865	
	4h	(b) Bubuowly Indigo Factory, Supuhee,		200·0
348	3e	**Vernon County**, Wisconsin, U.S.A. .	March 26, 1865	52·1
349	3e	**Gopalpur**, Jessore, India . . .	May 23, 1865	147·0
350	3d	**Dundrum**, Tipperary, Ireland . .	Aug. 12, 1865	245·0
351	3e	**Aumale**, Constantine, Algeria . .	Aug. 25, 1865	9·1
352	4o	**Sherghotty**, near Gya, Behar, India .	Aug. 25, 1865	118·8
353	4n	**Muddoor**, Mysore, India . . .	Sept. 21, 1865	407·3
354	3c	**Udipi** (Yedabettu), South Canara, India.	April 1866	3,306·0
355	3e	**Pokhra**, near Bustee, Goruckpur, India .	May 27, 1866	45·9
356	3e	**St. Mesmin**, Aube, France. . .	May 30, 1866	109·8
357	3c,4d 4h,4n	**Knyahinya**, near Nagy-Berezna, Hungary.	June 9, 1866	13,053·0
358	3e	**Jamkheir**, Ahmednuggur, Bombay .	Oct. 5, 1866	18·8

No.	Pane.	Name of Meteorite and Place of Fall.	Date of Fall.	Weight in grams.
359	3e	**Cangas de Onis**, Asturias, Spain .	Dec. 6, 1866	96·5
360	3e	**Khetrie** (Sankhoo, Phulee, &c.), Rajpootana, India.	Jan. 19, 1867	13·1
361	4o	**Tadjera**, near Guidjel, Setif, Algeria .	June 9, 1867	39·6
362	4e–g	**Pultusk** (Siedlce, Gostkóv, &c.), Poland.	Jan. 30, 1868	17,905·5
363	3f, 4d	**Daniel's Kuil**, Griqualand, South Africa.	March 20, 1868	449·5
364	3e	**Slavetic**, Agram, Croatia, Austria .	May 22, 1868	20·7
365	3e	**Ornans**, Doubs, France . . .	July 11, 1868	1,018·5
366	3f	**Sauguis**, St. Étienne, Basses-Pyrénées, France.	Sept. 7, 1868	15·8
367	3f	**Danville**, Morgan County, Alabama, U.S.A.	Nov. 27, 1868	27·2
368	3n	**Frankfort** (4 miles S. of), Franklin County, Alabama, U.S.A. [India.	Dec. 5, 1868	32·0
369	3e	**Moteeka Nugla**, Ghoordha, Bhurtpur,	Dec. 22, 1868	407·9
370	3c, 4d	**Hessle**, near Upsala, Sweden . .	Jan. 1, 1869	910·4
371	3f	**Krähenberg**, Zweibrücken, Rhenish Bavaria.	May 5, 1869	2·8
372	3d	**Cléguérec** (Kernouve), Morbihan, France.	May 22, 1869	9,346·8
373	3f	**Tjabé**, Padangan, Java . . .	Sept. 19, 1869	134·5
374	3f	**Stewart County** (12 miles S.W. of Lumpkin), Georgia, U.S.A.	Oct. 6, 1869	17·4
375	3n	**Ibbenbühren**, Westphalia, Prussia .	June 17, 1870	3·0
376	3f	**Cabeza de Mayo**, Murcia, Spain .	Aug. 18, 1870	3·4
377	4o	**Roda** (4 miles from), Huesca, Spain .	Spring 1871	7·7
378	3f	**Searsmont**, Waldo County, Maine, U.S.A.	May 21, 1871	51·5
379	3f	**Laborel**, Drôme, France.	June 14, 1871	291.5
380	3g	**Bandong**, Java.	Dec. 10, 1871	14·0
381	4d	**Dyalpur**, Sultanpur, Oude, India. .	May 8, 1872	269·8
382	3g	**Tennassilm**, Esthland, Russia . .	June 28, 1872	15·8
383	3g	**Lancé:** { Authon and Lancé, Vendôme, Loir-et-Cher, France. } July 23, 1872		332·9
384	4o	**Orvinio**, near Rome, Italy . . .	Aug. 31, 1872	62·8
385	3e	**Jhung**, Punjaub, India . . .	June 1873	1,984·0
386	3f	**Khairpur**, 35 miles east of Bhawalpur, India.	Sept. 23, 1873	2,991·0
387	3h	**Santa Barbara**, Rio Grande do Sul, Brazil.	Sept. 26, 1873	1·7
388	3h	**Aleppo**, Syria	Fell about 1873	67·0
389	3h	**Sevrukovo**, near Belgorod, Kursk, Russia. [Carolina, U.S.A.	May 11, 1874	20·1
390	3h	**Nash County** (near Castalia), N.	May 14, 1874	29·4
391	3k	**Virba**, Vidin, Turkey . . .	May 20, 1874	38·5
392	3h	**Kerilis**, Mael Pestivien, Côtes-du-Nord, France. [U.S.A.	Nov. 26, 1874	74·9
393	3f	**West Liberty**, Iowa County, Iowa,	Feb. 12, 1875	3,780·0
394	3f	**Sitathali** (Nurrah), S.E. of Raepur, Central Provinces, India.	March 4, 1875	600·0

No.	Pane.	Name of Meteorite and Place of Fall.	Date of Fall.	Weight in grams.
395	4d	**Zsadány**, Temeser Banat, Hungary .	March 31, 1875	25·2
396	3n	**Nageria**, Fathabad, Agra, India . .	April 24, 1875	13·5
397	3f	**Mornans**, Bourdeaux, Drôme, France .	Sept. 1875	975·0
398	4n	**Judesegeri**, Kadaba Taluk, Mysore, India.	Feb. 16, 1876	135·1
399	3h	**Vavilovka**, Kherson, Russia . .	June 19, 1876	10·3
400	3g	**Ställdalen**, Nya Kopparberg, Örebro, Sweden.	June 28, 1876	1,563·0
401	3k	**Rochester**, Fulton County, Indiana, U.S.A. [U.S.A.	Dec. 21, 1876	8·5
402	3k	**Warrenton**, Warren County, Missouri,	Jan. 3, 1877	82·5
403	3k	**Cynthiana** (9 miles from), Harrison County, Kentucky, U.S.A.	Jan. 23, 1877	154·8
404	3k	**Hungen**, Hesse, Germany . . .	May 17, 1877	5·4
405	3k	**Jodzie**, Ponevej, Kovno, Russia .	June 17, 1877	1·6
406	3h	**Soko-Banja**, N.E. of Alexinatz, Servia.	Oct. 13, 1877	1,975·0
407	3h	**Cronstadt**, Orange River Free State, S. Africa.	Nov. 19, 1877	1,226·6
408	3l	**Bhagur**, India	Nov. 27, 1877	10·5
409	3k	**Tieschitz**, Prerau, Moravia. . .	July 15, 1878	17·3
410	3h	**Dandapur**, Goruckpur, India . .	Sept. 5, 1878	2,245·0
411	3k	**Rakovka**, Tula, Russia . . .	Nov. 20, 1878	375·0
412	3l	**La Bécasse**, Dun le Poëlier, Indre, France.	Jan. 31, 1879	19·5
413	4o	**Angra dos Reis**, Rio de Janeiro, Brazil.	Jan. 1879	6·3
414	3l	**Itapicuru-mirim**, Maranhão, Brazil .	March 1879	6·4
415	3l	**Gnadenfrei**, Prussian Silesia . .	May 17, 1879	54·1
416	3m	**Nagaya**, Entre Rios, Argentine Republic.	July 1, 1879	7·0
417	3l	**Kalambi**, Bombay, India . . .	Nov. 4, 1879	28·0
418	3l	**Tomatlan**, Jalisco, Mexico . .	Sept. 17, 1879	135·7
419	3l	**Middlesbrough**, Yorkshire . .	March 14, 1881	25·6
420	3l	**Pacula**, Jacala, Hidalgo, Mexico. .	June 18, 1881	28·0
421	3l	**Gross-Liebenthal**, 12 miles S.S.W. of Odessa, Russia.	Nov. 19, 1881	62·5
422	2p 3l,4d	**Mocs**, Kolos, Transylvania . . .	Feb. 3, 1882	14,510·0
423	3l	**Fukutomi**, Hizen, Japan . . .	March 19, 1882	4·5
424	3n	**Pavlovka**, Balachev, Saratov, Russia .	Aug. 2, 1882	78·0
425	3l	**Pirgunje**, Dinagepur, India. . .	Aug. 29, 1882	734·0
426	3l	**Saint Caprais-de-Quinsac**, Gironde, France	Jan. 28, 1883	9·2
427	3m	**Alfianello**, Brescia, Italy . . .	Feb. 16, 1883	2,515·0
428	3l	**Pirthalla**, Hissar District, Punjaub, India.	Feb. 9, 1884	427·0
429	3l	**Djati-Pengilon**, Java . . .	March 19, 1884	469·0
430	3m	**Tysnes** island, Hardanger Fiord, Norway.	May 20, 1884	896·0
431	3l	**Chandpur**, 5 miles N.W. of Mainpuri, North-West Provinces, India.	April 6, 1885	490·5
432	3m	**Nammianthal**, South Arcot, Madras, India.	Jan. 27, 1886	1,623·0
433	3l	**Assisi**, Perugia, Italy	May 24, 1886	152·0

F

No.	Pane.	Name of Meteorite and Place of Fall.	Date of Fall.	Weight in grams.
434	3m	**Alatyr**, Karamzinka, Petrovka, Nijni Novgorod, Russia.	Sept. 4, 1886	22·0
435	3p, 3m	**Yenshigahara**, Kita-isa, Kagoshima, Satsuma, Kinsiu, Japan.	Nov. 10, 1886	31,030·0
436	3m	**Bielokrynitschie**, Zaslavl, Volhynia, Russia.	Jan. 1, 1887	54·0
437	3m	**Lalitpur**, North-West Provinces, India.	April 7, 1887	82·2
438	3m	**Tabory**, Ochansk, Perm, Russia . .	Aug. 30, 1887	1,222·0
439	3n	**Lundsgård**, Ljungby, Sweden . .	April 3, 1889	214·0
440	3n	**Mighei**, Olviopol, Elizabetgrad, Kherson, South Russia.	June 18, 1889	87·2
441	3n	**Jelica**, Servia	Dec. 1, 1889	1,879·0
442	3n	**Collescipoli**, Terni, Italy . . .	Feb. 3, 1890	105·0
443	3m	**Baldohn**, Misshof, Courland, Russia .	April 10, 1890	134·0
444	3n	**Winnebago County**, Iowa, U.S.A. .	May 2, 1890	2,560·0
445	3n	**Kahangarai**, Tirupatúr, Salem, Madras, India.	June 4, 1890	122·0
446	3n	**Washington**, Washington County, Kansas, U.S.A.	June 25, 1890	802·0
447	3m	**Indarh**, Elissavetpol, Transcaucasia .	April 7, 1891	42·9
448	3m	**Cross Roads**, Wilson County, N. Carolina, U.S.A.	May 24, 1892	11·8
449	3m	**Bath**, S. Dakota, U.S.A. . . .	Aug. 29, 1892	2,119·0
450	3m	**Bherai**, Junagadh, Kathiawar, Bombay	April 28, 1893	17·4
451	3m	**Beaver Creek**, West Kootenai District, British Columbia.	May 26, 1893	685·5
452	3m	**Zabrodje**, Wilna, Russia . . .	Sept. 22, 1893	3·0
453	3m	**Fisher**, Polk County, Minnesota, U.S.A.	April 9, 1894	602·0
454	3m	**Bori**, Badnúr, Betul District, Central Provinces, India.	May 9, 1894	1,270·0
455	3m	**Bishunpur** (and Parjabatpur), Mirzapur District, North-West Provinces, India.	April 26, 1895	393·5
456	3m	**Ambapur Nagla**, Sikandra Rao Tahsil, Aligarh District, North-West Provinces, India.	May 27, 1895	1,075·5
457	3m	**Madrid**, Spain	Feb. 10, 1896	
458	3m	**Lesves**, Nemur, Belgium . . .	April 13, 1896	

B. FALL NOT RECORDED.
[Arranged geographically.]

No.	Pane.	Name of Meteorite and Place of Find.	Report of Find.	Weight in grams.
459	3o	**Mainz**, Hesse, Germany. Described in 1857 by Scelheim : it had been turned up by a plough some years before.	Jahrb. d. Ver. für Naturk. im Nassau, 1857, p. 405.	33·6

No.	Pane.	Name of Meteorite and Place of Find.	Report of Find.	Weight in grams.
460	3o	**Oczeretna**, Lipovitz, Kiev, Russia. Found in the summer of 1871.		117·2
461	3o	**Assam**, India. Found in 1846 in the refuse of the "Coal and Iron Committee's" collections, probably obtained from Assam.	Proc. Asiatic Soc. Bengal, June, 1846, pp. xlvi, lxxvi.	538·7
462	4h	**Goalpara**, Assam, India. Found among some specimens obtained from the neighbourhood of Goalpara : described by Haidinger in 1869.	Wien. Akad. Ber. 1869, vol. 59, part 2, p. 665.	1,187·0
463	Sec.	**Barratta**, Deniliquin, New South Wales. One person thought he saw it fall in the month of May, about 1860 : another reports that he saw it lying on the ground in 1845.	Trans. Roy. Soc. of New South Wales, 1872, vol. 6, p. 97.	Sections only.
464	3o	**Makariwa**, Invercargill, New Zealand. Found in clay, about 2½ ft. from the surface, in 1879 : described by Ulrich and L. F. in 1893–4.	Proc. Roy. Soc., 1893, vol. 53, p. 54: Mineralog. Magazine, 1894, vol. 10, p. 287.	62·8
465	3o	**Tomhannock Creek**, Rensselaer County, New York, U.S.A. Found about the year 1863 : described by Bailey in 1887 : Brezina points out a close likeness of this stone and also of "Yorktown" to those of West Liberty.	Amer. Jour. Sc. 1887, ser. 3, vol. 34, p. 60 : Ann. d. k. k. Naturh. Hofmus. Wien, 1896, vol. 10, p. 251.	17·2
466	3o	**Morristown**, Hamblen County, Tennessee, U.S.A. Found in 1887 : described by Eakins in 1893.	Amer. Jour. Sc. 1893, ser. 3, vol. 46, p. 283.	561·7
467	3o	**Waconda**, Mitchell County, Kansas, U.S.A. Found in 1873 in the grass, upon the slope of a ravine : described by Shepard and by Patrick in 1876.	Amer. Jour. Sc. 1876, ser. 3, vol. 11, p. 473 : Trans. Kansas Ac. Sc. 1876, vol. 5, p. 12.	467·5
468	3o	**Prairie Dog Creek**, Decatur County, Kansas, U.S.A.	Tschermak's Min. und Petrog. Mittheil. Wien, 1894–5, vol. 14, p. 471.	525·0
469	3o	**Long Island**, Phillips County, Kansas, U.S.A. This and the preceding were reported and described by Weinschenk in 1895.	*Ibid.*	
470	3o	**Utah**, U.S.A. Found in 1869 on the open prairie between Salt Lake City and Echo, Utah : described by Dana and Penfield in 1886.	Amer. Jour. Sc. 1886, ser. 3, vol. 32, p. 226.	4·7
471	3o	**MacKinney**, Collin County, Texas, U.S.A.		290·0

F 2

No.	Pane.	Name of Meteorite and Place of Find.	Report of Find.	Weight in grams.
472	3o	**Bluff,** 3 miles S. W. of La Grange, Fayette County, Texas. Found about 1878, and described by Whitfield and Merrill in 1888.	Amer. Jour. Sc. 1888, ser. 3, vol. 36, p. 113.	12,700·0
473	3o	**Pipe Creek,** Bandera County, Texas, U.S.A. Found in 1887: described by Ledoux in 1888–9.	Trans. of New York Ac. of Sc., 1888–9, vol. 8, p. 186.	821·0
474	3o	**The Lutschaunig Stone,** Atacama, Chili.	Mineralog. Magaz. 1889, vol. 8, p. 234.	92·0
475	3o	**Carcote,** Atacama, Chili, S. America. Known since 1888: described by Sandberger in 1889.	Jahrb. f. Min.,1889, vol. 2, p. 173.	2·7
476	3o	**Minas Geraes** (?), Brazil. Found without label among specimens which may have been brought from Minas Geraes: mentioned by Derby in 1888.	Revista do Observatorio, Rio de Janeiro, 1888.	3·6

APPENDIX A.

NATIVE IRON (terrestrial).

(Pane 4m).

Name of Iron and Place of Find.	Report of Find.	Weight in grams.
Sowallick Mountain, West Greenland (Ross's iron). Two knives with bone handles given to Captain John Ross in 1818 by the Esquimaux of Prince Regent's Bay : one of them is that figured by Ross on page 102 of his work. According to the Esquimaux, the iron had been obtained from a neighbouring mountain called Sowallick.	Voyage of Discovery, &c., by Captain John Ross. London, 1819.	
Upernavik, West Greenland (Kane's iron). Dr. Kane saw walrus-lances tipped with iron in the possession of the Esquimaux who visited the brig in its winter quarters at Rensselaer Harbour, Smith Sound, in 1854. He learned afterwards that the iron was obtained in traffic from the more southern tribes. Perhaps it was got from Sowallick Mountain.	Arctic Explorations, by Dr. E. K. Kane. Philadelphia, 1856, vol. 1, p. 206.	1·4
Niakornak, Jakobshavn District, West Greenland (Rink's iron). Part of a lump obtained (1848–50) by Dr. Rink from a Greenlander who lived at Niakornak : it had been found not far from his home, lying loose on a pebble-strewn plain near the coast.	Oversigt over det koniglike danske vidensk. selsk. forh. 1854, p. 1.	2,023·0
Jakobshavn, West Greenland (The Pfaff-Öberg iron). Part of a lump given by Dr. Pfaff of Jakobshavn to Dr. Öberg in 1870 : it was said to have been found in the neighbourhood (perhaps near Niakornak).	Geological Magazine, 1872, vol. 9, p. 520.	290·4
Ovifak, Disko Island, West Greenland. Found by Nordenskiöld in 1870.	Geological Magazine, 1872, vol. 9, p. 460.	90,300·0
New Zealand (Jackson's Bay). Found in 1885, and described by Skey in the same year (Awaruite).	Trans. and Proc. of New Zealand Institute, 1885, vol. 18, p. 401.	4·7

APPENDIX B.

PSEUDO-METEORITES (Drawer).

Aachen, Rhenish Prussia.
Braunfels, Coblenz.
Campbell County, Tennessee, U.S.A.
Canaan, Connecticut, U.S.A.
Clough, Antrim, Ireland.
Collina di Brianza, Milan, Italy.
Concord, New Hampshire, U.S.A.
Eisenberg, Saxon Altenburg.
Gross-Kamsdorf, Saxony.
Heidelberg, Germany.
Hommoney Creek, Buncombe County, N. Carolina, U.S.A.
Igast, Livland, Russia.
Kamtschatka, Asiatic Russia.
Leadhills, Lanarkshire, Scotland.
Long Creek, Jefferson County, New York, U.S.A.
Magdeburg, Prussia.
Minsk (Mozyr), Russia.
New Haven, Connecticut, U.S.A.
Nöbdenitz, Saxon Altenburg.
Richland, S. Carolina, U.S.A.
Rutherfordton, N. Carolina, U.S.A.
Sterlitamak, Russia.
Voigtland, Saxony.
Waterloo, New York, U.S.A.
Yafaee Mountains, Arabia.

LIST OF THE CASTS OF METEORITES.

Meteorites are generally represented in collections by mere fragments of the original specimens, which often fail to give any idea of the original size and shape. Before division of a specimen a cast of it is sometimes prepared, and a representation of the size and shape is thus preserved.

Casts of most of the following meteorites are exhibited in the lower parts of the cases :—

Akburpur.
Assisi.
Barranca Blanca.
Babb's Mill.
Barratta
Beuste.
Bingera.
Bithur.
Braunau.
Breitenbach.
Buschhof.
Bustee.
Butsura.
Cabin Creek.
Cachiyuyal.
Charlotte.
Chulafinnee.
Cronstadt.
Daniel's Kuil.
Dolgovoli.
Dundrum.
Durala.
Goalpara.
Gopalpur.
Ibbenbühren.
Jelica.
Jhung.
Kaee.
Khiragurh.
Klein-Menow.
Launton.
Lick Creek.
Linum.

Mazapil.
Mhow.
Middlesbrough.
Mooresfort.
Mouza Khowrna.
Nagy-Diwina.
Nash County.
Nedagolla.
Nejed.
Nellore.
Nerft.
Newstead.
New Zealand.
Obernkirchen.
Ogi.
Ovifak.
Parnallee.
Petersburg.
Pillistfer.
Pulsora.
Rancho de la Pila.
Rittersgrün.
Roebourne.
Rowton.
St. Denis Westrem.
Sarepta.
Segowlie.
Shytal.
Sitathali.
Ski.
Udipi.
Virba.
West Liberty.

The Trustees possess moulds of those meteorites in the preceding list of which the names are printed in italics, and casts may be obtained on payment of the necessary expenses. Applications should be made in writing to the formatori, D. Brucciani & Co., 40 Russell Street, Covent Garden, London.

INDEX

TO THE METEORITES REPRESENTED IN THE COLLECTION.

The names adopted for the meteorites are printed in thick type: the other names are synonyms.
The numbers correspond with those of the first column of the meteorite list.

www.ingramcontent.com/pod-product-compliance
Lightning Source LLC
Chambersburg PA
CBHW020030030726
47499CB00007B/2350